Proceedings of the Thirty-third
Annual Biology Colloquium

The Annual Biology Colloquium

YEAR, THEME, AND LEADER

1939. *Recent Advances in Biological Science.* Charles Atwood Kofoid

1940. *Ecology.* Homer LeRoy Shantz

1941. *Growth and Metabolism.* Cornelius Bernardus van Niel

1942. *The Biologist in a World at War.* William Broadbeck Herns

1943. *Contributions of Biological Science to Victory.* August Leroy Strand

1944. *Genetics and the Integration of Biological Sciences.* George Wells Beadle

1945. (Colloquium cancelled)

1946. *Aquatic Biology.* Robert C. Miller

1947. *Biogeography.* Ernst Antevs

1948. *Nutrition.* Robert R. Williams

1949. *Radioisotopes in Biology.* Eugene M. K. Geiling

1950. *Viruses.* W. M. Stanley

1951. *Effects of Atomic Radiations.* Curt Stern

1952. *Conservation.* Stanley A. Gain

1953. *Antibiotics.* Wayne W. Umbreit

1954. *Cellular Biology.* Daniel Mazia

1955. *Biological Systematics.* Ernst Mayr

1956. *Proteins.* Henry Borsook

1957. *Arctic Biology.* Ira Loren Wiggins

1958. *Photobiology.* F. W. Went

1959. *Marine Biology.* Dixy Lee Ray

1960. *Microbial Genetics.* Aaron Novick

1961. *Physiology of Reproduction.* Frederick L. Hisaw

1962. *Insect Physiology.* Dietrich Bodenstein

1963. *Space Biology.* Allan H. Brown

1964. *Microbiology and Soil Fertility.* O. N. Allen

1965. *Host-Parasite Relationships.* Justus F. Mueller

1966. *Animal Orientation and Navigation.* Arthur Hasler

1967. *Biometeorology.* David M. Gates

1968. *Biochemical Coevolution.* Paul R. Erlich

1969. *Biological Ultrastructure: The Origin of Cell Organelles.* John H. Luft

1970. *Ecosystem Structure and Function.* Eugene P. Odum

1971. *The Biology of Behavior.* Bernard W. Agranoff

1972. *The Biology of the Oceanic Pacific.* John A. McGowan

1973. *The Biology of Tumor Viruses.* Joseph W. Beard

The Biology of the Oceanic Pacific

Proceedings of the Thirty-third
Annual Biology Colloquium

Edited by CHARLES B. MILLER

Corvallis:
OREGON STATE UNIVERSITY PRESS

Library of Congress Cataloging in Publication Data

Biology Colloquium, 33d, Oregon State University, 1972.
    The biology of the oceanic Pacific.

    (Proceedings of the Annual Biology Colloquium, 33d)
    Includes bibliographies.
    1. Marine biology—Pacific Ocean—Congresses.
I.   Miller, Charles B., 1940-   ed.   II.   Title.
III.   Series: Biology Colloquium, Oregon State Uni-
versity. Annual Biology Colloquium proceedings, 33d.
[DNLM: 1.   Ecology—Congresses.   2.   Marine biology—
Congresses.   3.   Oceanography—Congresses.   W3BI554
33d 1972 / GC11 B615 1972b]
QH301.B43 33d [QH95] 574′.08s [574.92′5] 74-4300
ISBN 0-87071-172-5

Preface

THE YEARS 1872-1876 mark a departure in our view of the oceans and their life. The *Challenger* Expedition began the revelation that the far reaches of the ocean are substantially different as environments from coastal waters. It is appropriate during the centennial of those years to assemble what we have learned about life in the oceans and to restate it clearly. The 1972 Biology Colloquium at Oregon State University was one attempt to do that.

For historical reasons most of the biologists who have studied the oceanic regions have been systematists or ecologists. The interplay between these disciplines has produced a great share of our knowledge. This is reviewed for pelagic environments by John A. McGowan (The Nature of Oceanic Ecosystems), and for benthic environments by Robert R. Hessler (The Structure of Deep Benthic Communities from Central Oceanic Waters). In recent decades the trophodynamic viewpoint has been applied to oceanic studies by Riley, Strickland, and a host of others. The results are reviewed by Timothy R. Parsons and Bodo R. de Lange Boom (Control of Ecosystem Processes in the Sea) and by Bruce W. Frost (Feeding Processes at Lower Trophic Levels in Pelagic Communities). The practical importance of the far ocean reaches for purposes other than warfare and shipping has come to be recognized only rather recently. The biological aspects of this include the oceanic fisheries, whose prospects are reviewed by Brian Rothschild (Fishery Potential from the Oceanic Regions). The molecular adaptations that allow survival in the oceanic environment, particularly at great depths, are a vast reservoir of potential research projects only very recently tapped. The review by Peter W. Hochachka (Enzymatic Adaptations to Deep Sea Life) discusses the results.

Together these papers are a review of deep-sea biology. No pretense of completeness is made, however, and to a degree the papers are simply a small sampling of viewpoints in oceanic biology. One other viewpoint is presented: Joel Hedgpeth has provided us with an his-

5

142075

torical review of Pacific biology. It is partly a personal review, and I wish he would repeat the performance at book length.

I would like to thank the other members of the 1972 Biology Colloquium Committee (John V. Byrne, Robert Holton, James McCauley, and J. Kenneth Munford) for their help with the project, and I would like to thank all of the sponsoring organizations that made it possible.

CHARLES B. MILLER
Oregon State University

Contents

The colloquium leader, John A. McGowan, at work at sea.

The Nature of Oceanic Ecosystems

JOHN A. McGOWAN
Scripps Institution of Oceanography
La Jolla, California

ONE OF THE MAJOR objectives of the *Challenger* Expedition was to define and describe the flora and fauna of the open ocean. The expedition was an enormous success in meeting this aim. For perhaps the next fifty years other expeditions with very similar objectives carried on this work. However, about in the 1920's, it became quite apparent that the fisheries resources of the seas were not limitless. Various agencies began the difficult and tedious studies of the factors influencing population variability of marine organisms, primarily commercial fishes, and thus shifted the emphasis in biological oceanography away from exploratory studies of the open ocean. In the late 1940's and early 1950's it was descriptive physical oceanographers who carried on and expanded the biological sampling of the oceanic Pacific. This came about because they felt a need for better data on the distribution of the physical properties, and some of them, at least, were willing to include biological sampling as part of their routine of observation and measurement. This resulted in the accumulation of a large library of macrozooplankton samples and a somewhat smaller set of nekton samples caught by the Isaacs-Kidd midwater trawl. One of the values of these collections is due to the physical oceanographers' design of expedition tracks and selection of sampling locations. These were intended to provide data for the definition of water masses and current systems and their boundaries and to help understand large-scale mixing processes. As it turned out, these objectives were eminently well suited to and compatible with the study of quantitative oceanic biogeography.

I have previously listed (McGowan, 1971) what I believe to be the objectives of many biogeographers. These are: (1) to determine what species are present; (2) to describe, quantitatively, their patterns

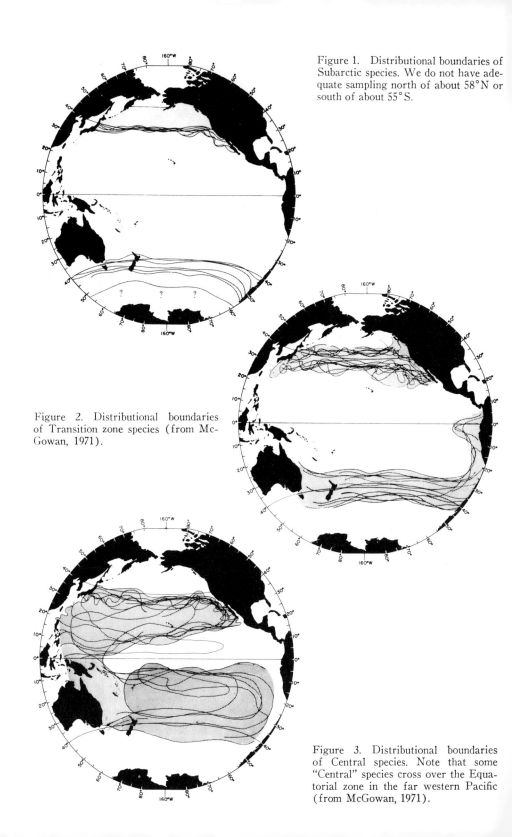

Figure 1. Distributional boundaries of "Subarctic species. We do not have adequate sampling north of about 58°N or south of about 55°S.

Figure 2. Distributional boundaries of Transition zone species (from McGowan, 1971).

Figure 3. Distributional boundaries of Central species. Note that some "Central" species cross over the Equatorial zone in the far western Pacific (from McGowan, 1971).

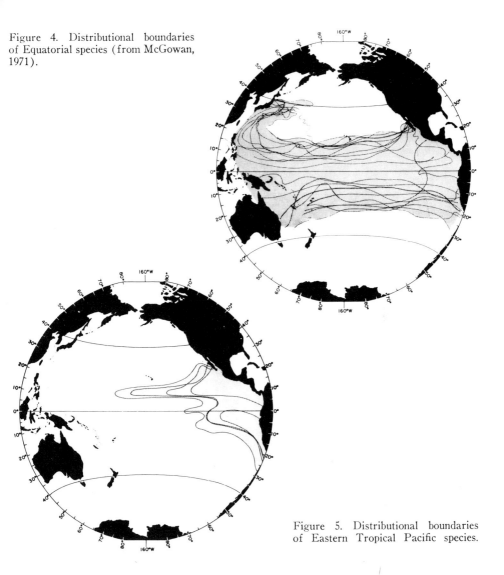

Figure 4. Distributional boundaries of Equatorial species (from McGowan, 1971).

Figure 5. Distributional boundaries of Eastern Tropical Pacific species.

of abundance; (3) to understand what maintains the patterns; (4) to determine how and why the patterns developed; and (5) to describe and delineate the communities. I would now add to this list: (6) to determine how these community-ecosystems are structured and how they function. The first two objectives are essentially those of the *Challenger,* and we now have a reasonably good knowledge of species and pattern for several groups of macrozooplankton and "micro-nekton" and are beginning to understand phytoplankton and "macro-nekton." The maintenance of pattern is not very well understood but seems to be involved with the major climatic and oceanographic processes that lead to the structuring of water masses. Determining how these patterns arose has proven to be very difficult, but the process seems to be involved with paleocirculation systems interacting with mechanism(s) of speciation.

The survey cruises and expeditions from the *Challenger* up to the beginning of the 1960's were appropriate for the purposes of ecologically oriented biogeography and such work is still badly needed. However, the type of sampling regime imposed by the necessity of covering great distances was not well suited to providing the type of data required to meet the latter two objectives of defining and understanding community structures and function. Therefore, during the 1960's and continuing at present, much oceanic work is being done as time series in rather small circumscribed locales which we have reason to believe are "representative" of much larger habitats. This work is characterized by intensive replication of sampling, measurement, and experiment. The selection of these locales was based on our previous knowledge of pattern. Although this work has only begun, it is useful at this point to attempt to review what we know or think we know of the character of oceanic ecosystems.

The patterns

Figures 1 through 5 illustrate the patterns of distribution of a variety of planktonic and nektonic species. These show the maximum areal ranges reported for these species. The organisms included in preparing these illustrations are: euphausiids, chaetognaths, pteropods, foraminifera, some genera of copepods, ommastrephid squid, tuna, and salmon. Other more recent information shows that myctophid fishes, sergestid shrimp, and diatoms all tend to fit these patterns. In general all species of oceanic organisms for which we have adequate biogeographic data were used in the construction of these maps. It is evident that there are rather few basic patterns; they are: Subarctic and Subantarctic (Fig. 1), North and South Transition zones (Fig. 2), North and South Central (Fig. 3), Equatorial (Fig. 4), Eastern Tropical Pacific (Fig. 5), and Warm Water "Cosmopolites."

In addition to these patterns there is a distinct fauna in the Southern Ocean (the Antarctic) and there are endemic species of macrozooplankton and fish in the California and Humboldt-Peru Current systems. These latter endemics are apparently few in number.

It is obvious that the boundaries of these major habitats overlap broadly. However, there are two other methods of presenting this information that indicate a strong quantitative separation of the faunas.

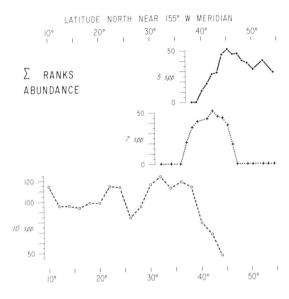

Σ RANKS

ABUNDANCE

Figure 6. The summed ranks of abundance of some Sub-arctic (upper graph), Transition (center), and Warm Water Cosmopolite (lower) species along 155° W (from McGowan, 1971).

Figure 6 is the sums of the ranks of abundance for 22 species of Sub-arctic, Transition Zone, and Warm Water Cosmopolite organisms for which we have estimates of abundance along 155° W. It shows that while there is some admixture of these species, their peaks of abundance are fairly well separated. A visual inspection of the patterns of abundance of these and many other species (see for examples Brinton, 1962 and McGowan, 1971) will also convey a similar picture, namely that many species (except the Warm Water Cosmopolites) are least abundant near the peripheries of their ranges. The degree of "separation" of these faunas may also be seen in Figures 7 through 12. In these maps I have tried to estimate the percent of a fauna that is present. The information for these maps came from Figures 1 through 5 and more recent information. In all cases the darkest crosshatching indicates that a "pure" fauna was present. But in the zone of intermediate cross-hatching about 40 percent of the species have dropped out and in the lightest area about 70 percent. The contour intervals, then, are in order of decreasing fidelity to the central "core." It does *not* mean that in every appropriate sample from the dark area all species of that fauna will be present, but merely that the contour interval includes the range of all species that we know are indigenous to the areas. This latter point may be well illustrated by Figure 13, in which only the samples containing all members of recurrent groups of zooplankton species are shown (Fager and McGowan, 1963). This illustrates that species which have a "high" frequency of co-occurrence in general tend to

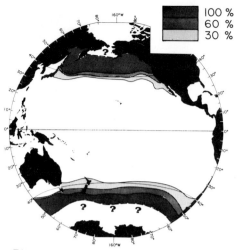

Figure 7. The 100 percent level includes all Subarctic species but 60 percent of these have a somewhat broader range, particularly in the California Current and 30 percent have an even broader range. Question marks indicate a lack of adequate, quantitative sampling.

ESTIMATED PERCENT OF SUBARCTIC
OR SUBANTARCTIC FAUNA PRESENT

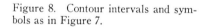

Figure 8. Contour intervals and symbols as in Figure 7.

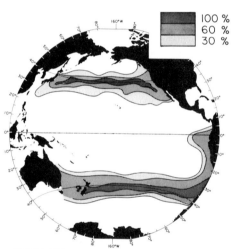

ESTIMATED PERCENT OF TRANSITION
ZONE FAUNA PRESENT

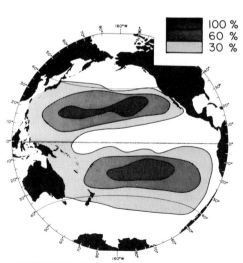

ESTIMATED PERCENT OF CENTRAL
FAUNA PRESENT

Figure 9. Contour intervals and symbols as in Figure 7.

Figure 10. Contour intervals and symbols as in Figure 7.

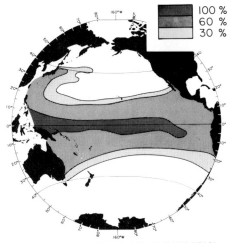

ESTIMATED PERCENT OF EQUATORIAL
FAUNA PRESENT

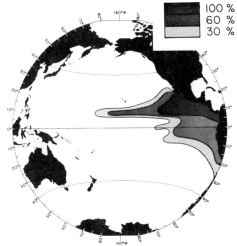

ESTIMATED PERCENT OF EASTERN
TROPICAL PACIFIC FAUNA PRESENT

Figure 11. Contour intervals and symbols as in Figure 7.

Figure 12. Contour intervals and symbols as in Figure 7.

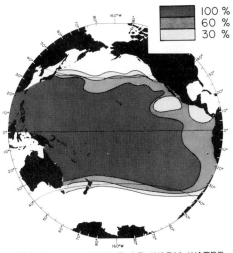

ESTIMATED PERCENT OF WARM WATER
"COSMOPOLITES" FAUNA PRESENT

occur as a unit (i.e., all members present) in the same sample at many locales within the Subarctic habitat. The same is true of the Transition, Central, and Equatorial habitats.

One might ask: Do the species of these habitats tend to "agree" on where to be abundant and where to be relatively rare within their habitats? If they do, it may reasonably be concluded that the species are responding, by their abundance or rarity, in a similar manner to areal variations in their environment which, in turn, implies a considerable degree of co-adaptation. A concordance test of the ranks of abundance over the locales sampled is an appropriate test for this question. In all cases significant positive concordance was found (Fager and McGowan, 1963). Finally, we expect organized, highly "evolved" communities to be made up of species which have undergone a long period of evolutionary adaptation together. In such a situation many competitive interactions should have taken place already and we have the "results" of that competition. If this is so, then we should see considerable discrimination against closely related (i.e., functionally similar) species occuring "together" in the same recurrent group. Such is the case for the four taxonomic categories studied (Fager and McGowan, 1963).

Thus there is a reasonable amount of evidence that the major oceanic biotic provinces of the north and to a lesser degree the south Pacific Ocean have been mapped and identified. There is strong circumstantial evidence that these provinces are ecosystems. What may we say about the character of these ecosystems?

Ecosystem characteristics

1. *They are large and few in number.* There are eight basic patterns shown in Figure 14. Six of these are made up to a considerable degree of species with amphitropical distributions—that is, the Subarctic and Subantarctic have many species in common as do the North and South Transition and North and South Central systems. Although these areas may share the same species, it is by no means certain at the present time that their communities are structured in the same way. One of the objectives of current oceanic studies is to find out if this is so.

These areas may be considered the oceanic equivalents of Dice's "Biotic Provinces" (1952, pp. 443-447). Although I have chosen to define their boundaries differently than Dice, the areas are very similar in many other properties. Dice identified 24 such provinces on the North American continent, while there are only eight in the entire pelagic zone of the Pacific. The Warm Water Cosmopolites cannot be considered to form a province in the same sense as the others, although

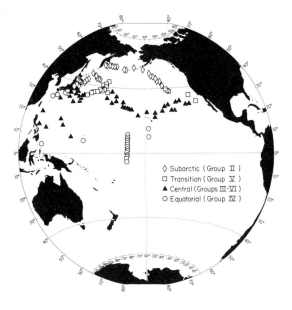

◊ Subarctic (Group Ⅱ)
□ Transition (Group Ⅴ)
▲ Central (Groups Ⅲ-Ⅵ)
○ Equatorial (Group Ⅳ)

Figure 13. The occurrence of samples containing n-members of recurrent groups of zooplankton species. The Warm Water Cosmopolites (Group I) are not shown on this chart as they occurred mostly in the Kuroskio and its extension. The blank areas in this chart within the biotic provinces shown in Figures 7 through 12 are due to lack of adequate sampling for recurrent group analysis (after Fager and McGowan, 1963).

Figure 14. The patterns of the basic (100% "core" regions) biotic provinces of the oceanic Pacific.

they may form a distinct recurrent group and a pattern in the core of the Kuroshio and its extension.

2. *They are semiclosed systems* but with a large amount of mixing along their outer peripheries (i.e., the areas of about 30% in Figures 7 through 12). Six of the basic patterns shown in Figure 14 are in areas where the circulation of water within the upper 1,000 m or so shows a strong tendency to recirculate. For example, the Subarctic province very nicely spans the area of the cyclonic circulation of the Subarctic gyre. The North and South Central provinces are near the geographical centers of the large anticyclonic gyres. Although neither the Eastern Tropical Pacific nor the Equatorial provinces are in gyres, there is a large amount of recirculation of water due to countercurrent systems. It is the two Transition provinces, north and south, that appear to be lacking physical evidence for being semiclosed systems. I have presented some (admittedly weak) evidence (McGowan, 1971) for occasional return, east to west, flow in the north based on the calculations and interpretations of Dodimead, Favorite, and Hirano (1963) but as yet there are no such data in the south.

3. *They are old.* The circulation of these areas is due to: (1) the size and shape of the Pacific Basin, (2) the direction of rotation of the earth, (3) the wind system resulting from the fact that higher latitudes are relatively cooler than the lower latitudes, and (4) variations in water density because of patterns in the ratio of precipitation to evaporation and in runoff.

Although there have been changes in sea level of the order of 100 m during the past hundred thousand years or more, the general shape of the Pacific Basin has remained the same for a very much longer time. Some of these driving forces for the circulation may have changed in their intensity over time, but it seems unlikely they have changed in sign. For example, Arrhenius (1963) has presented geochemical evidence that the Equatorial current system has maintained its present position since at least the Miocene. He further points out that apparently there have been changes in the speed of the currents. Because of the magnitude of the Equatorial current system and its close coupling with the rest of the circulation of the Pacific, it follows that the other major systems must also have more or less maintained their geographical position. Supporting evidence for this comes from Riedel and Funnell (1964, pp. 359-362), who have plotted the occurrences of calcareous-siliceous microfossils and point out that "it is reasonable to assume that they were deposited in a biologically productive region." They show that latitudinal boundaries of the productive Equatorial Current and the Subarctic systems have not changed since the beginning of the Pliocene. During the Oligocene and Eocene the equatorial boundaries seemed to have expanded somewhat but recent evidence (Riedel, pers. comm.) on sea floor spreading can explain this. Thus it seems that the basic circulation scheme of the Pacific Ocean has been the same for several tens of millions of years. It is highly probable that this is enough time for a considerable amount of evolutionary adaptation to take place. This could lead to adaptational succession and tend towards the establishment of climax communities in the semi-closed systems with species closely "tuned" to each other and to the physical environment, in other words, the establishment of ecosystems.

4. *They are highly tuned to climate but not to weather.* Within these biotic provinces a number of cruises have taken samples designed to reveal species abundance, community structure, nutrient distribution, and gross physical structure. During many of these cruises the sea state varied from dead calm to force six and on a few occasions force nine sea state was recorded. None of the variables measured was significantly correlated with these weather changes. On the other hand, Parsons and Le Brasseur (1968) have shown in the Subarctic a very marked, rapid, and regular response of productivity and zooplankton standing crops to seasonal changes in critical depth. The timing of these events is due to climate and is quite different than the timing of similar events in the Central system or the Eastern Tropical Pacific where a different seasonality is apparent. Further, there appear to be year-to-year variations in phytoplankton standing crops in the Central system which are coincident with rather subtle, large-scale climatic changes (Venrick, McGowan, and Mantyla, 1973).

5. *They tend to be monotonous.* Vertical sections of temperature, salinity, and nutrients within the upper few hundred meters and along both the major axes of some of these areas show very gentle or almost no horizontal gradients. There are no long-lasting "fronts" or major eddies. The phytoplankton and zooplankton standing crops tend to be uniform (or, rather, uniformly patchy) within seasons, and the relative proportions of zooplankton and small nektonic species tend to be monotonously similar from sample to sample. Other areas, the Equatorial for example, have rather strong north-south horizontal gradients of physical properties and thus are not homogeneous. In these cases, however, the "structural" features are very predictable. This relative homogeneity or predictability is probably brought about and maintained by recirculation and gyre-like circulation systems. Although macrozooplankton is probably as patchy in these systems as anywhere else, Miller (1970) has presented strong evidence for the ephemeral nature of small-scale species associations and thus these probably do not lead to the development of demes.

6. *There are quantitative and some qualitative differences in basic processes.* The Subarctic, Subantarctic, Eastern Tropical Pacific, and Equatorial ecosystems are highly eutrophic, but the two Central systems are oligotrophic. There are no easily interpretable estimates of primary productivity for either Transition province, but judging from the standing crops of both phytoplankton and zooplankton they may eventually be found to be between that of the Subarctic and Central. Because of the cyclonic circulation of the Subarctic gyre there is a generalized *upward* vertical movement of water amounting to some 20 ± 10 m/yr (Tully and Barber, 1960). This water comes from depths of at least 300 m and the movement is apparently rather widespread and continuous, varying with the speed of rotation of the gyre. However, because of variations in the mixed layer and euphotic zone depths, there is only a very brief period of high productivity in the late spring. This does not result in a phytoplankton "bloom" because of an almost immediate zooplankton grazing response.

Within the Equatorial province there are two large-scale physical features of biological importance. They exist as narrow, parallel, zonal bands. One of these is symmetrical about the equator. It is here that upwelling occurs because of the change in sign of the Coriolis force. Meridional (155° W), vertical sections show that the thermocline is very "weak" and the deeper isopleths of temperature, nitrate, phosphate, and silicate frequently "break the surface." It appears that this upwelled water comes from rather shallow depths, perhaps no more than 150 m. We know little of the rates of vertical movement of water or flux of nutrients to the euphotic zone in this band, but both phyto- and zooplankton standing crops tend to be higher here, at 0° (± 1°), than in the rest of the Equatorial system (King, 1954; Venrick, McGowan, and Mantyla, 1973). The second important physical feature of this province is located at about 9° N between the North Equatorial Current and the Equatorial Countercurrent. Here there is a "doming" of the isopleths of temperature and nutrients in a meridional section. These do not break the surface, and, since it is an area with very strong thermocline, it is unlikely that they ever do so. This thermocline is relatively shallow, and judging from the nutrient isopleths, fairly high levels of both nitrate and phosphate are present in the euphotic zone but below the thermocline. We do not, at present, understand the biological significance of this second feature, although there is some evidence that it has effects.

Between the zone of upwelling at 0° and the "ridge" (rather than a dome) at about 9° N is a zone of convergence. Thus the nutrients brought to the euphotic zone are mixed horizontally and vertically due to the current-countercurrent and the divergence-convergence-ridge systems. The system appears to be present always but at varying intensity. Therefore, while this province is anything but monotonous it is perhaps quite predictable.

Within the Eastern Tropical Pacific system nutrient input is more complex. It is due to two different processes. There is the classic, wind-driven, coastal upwelling, particularly in the Gulfs of Panama and Tehuantepec, and there is the geostrophic divergence of the Costa Rica Dome. The coastal upwelling is variable in intensity, depending on winds. The divergence of the Dome is variable in intensity and in geographic position, depending on the "strength" of the Equatorial Countercurrent (Wyrtki, 1966). Further, there is a very well developed oxygen minimum layer at relatively shallow depths. This feature alone has profound biological implications.

The two Central systems, with their anticyclonic circulation, are areas of generalized sinking. The major input of nutrients here apparently comes at a very brief period in midwinter when the depth of the mixed layer reaches the nutricline at 110 to 120 m and the nutrient-rich, deeper water "overturns" as in a lake.

Within the Subarctic primary productivity is light-limited with an "excess" of both nitrate and phosphate. Within the Eastern Tropical Pacific primary productivity is nitrate-limited with an "excess" of phosphate. Within the Central provinces both nitrate and phosphate are limiting. It is not clear at the present just what the limiting factors may be in the Equatorial province.

The Transition system remains an enigma. There is direct evidence for upward vertical mixing here (McGowan, 1971) and the theoretical work of Hidaka and Ogawa (summarized in Cushing, 1971) indicates that divergence should occur. Judging from the distribution of conservative properties (viz., salinity) this may be a relatively shallow ($<$200 m) process. We have little insight into either the rate-limiting factors or the regulation of the carrying capacity of this system.

7. *They are relatively undisturbed by man.* Several of the systems support rather large commercial fisheries; yellowfin tuna in the Eastern Tropical Pacific and Equatorial systems and pink salmon in the Subarctic are examples. This removal of top predators represents a disturbance by man, and we do not know the consequences to the rest of the community. But there do not, as yet, appear to be large changes in the zooplankton species structure of the Equatorial system from that of the pre-exploitation period of the early 1950's. This removal may be the only disturbing factor, for it is clear that the kind of habitat destruction and pre-emption that has so changed the land, lakes, estuaries, and even the littoral zone, has not taken place in these oceanic areas.

Further, most pollutants are present at relatively low levels and as yet no mass mortalities or major effects on reproduction, growth, or behavior have been reported. These may be the only large ecosystems still remaining in almost their "natural" state.

8. *Their basic organization does not differ from other ecosystems.* Nutrients enter from below, the sun's energy from above and these are transformed and transferred by plants, herbivores, omnivores, carnivores, and degraders. There are Eltonian pyramids of numbers and biomass and organisms can be lumped into Lindemanian trophic levels. Competition for "food" but not space occurs and predators may very well "regulate" their prey populations. Diversity is high in some systems and low in others, and some of these systems appear to be relatively more "stable" than others.

Thus these systems are very similar to those on land in many respects, but there are also differences, particularly quantitative differences. As compared to the land, nutrients are very dilute. All of the plants are very small and therefore so are the main herbivores. There is little or no cover; organisms are distributed three-dimensionally and are in almost constant movement. There is no direct analog to humus, peat, or forest litter, and degraders (bacteria, yeast, fungi) seem to be of minor importance. The particulate organic detritus of the ocean is, apparently, mostly of animal origin rather than of plant origin as it is on land. The "residence" time of organic detritus in the ocean is very short as compared to the land, and this detritus does not form an important habitat or physical niche for macroscopic animals. The ratio of dissolved to particulate organic matter is large.

In spite of their great antiquity there are relatively few species present in these systems as compared to similar-sized areas on land. There is a great deal of "leakage" along the western and eastern boundaries of all except the Central system. For example; of the 15×10^6 m^3/sec of water moving along the southern segment of the Subarctic cyclonic gyre, 10×10^6 m^3/sec "splits off," turns to the south, and forms part of the California Current. This "leaked" water is rich in nutrients and populations derived from the Subarctic ecosystem.

Unusual climatic variations are damped due to the high heat capacity of water and there are, of course, no problems associated with relative humidities. Plants do not "create" microclimates. Most of the recycling of essential plant nutrients (N and P) within the systems comes from animal excretion and not from the breakdown and decay of plant material, and the process is apparently very rapid. The primary productivity per unit area is generally much lower than most places on the land.

Thus these systems do differ in some very important ways from land systems and even from lakes. Because of this, it has been difficult to interpret oceanic ecosystems in terms of what we know about the land. Correlative to this, it seems unlikely to me that deterministic, mathematical ecosystem models of terrestrial or even lake systems can be readily applied to the ocean. Further, in view of the great differences between these oceanic systems (sinking vs. upwelling, for example), it would seem that this type of analysis would lack both generality and realism. That is, no one model could be applied to all systems without considerable modification. However, since it seems unlikely that, in the near future, we will develop the necessary facts and insight into the workings of these systems to achieve any reasonable degree of biological reality in models (Walsh, 1972), this perhaps is not an immediate, critical scientific problem. Some additional characteristics of these systems are shown in Table 1.

Ecotones, opportunists, and understanding ecosystems

Throughout the discussion of ecosystem characteristics, I have said very little of the Warm Water Cosmopolites. They obviously do not have many of the attributes described for the other faunas and we know very little about them. Perhaps they do form a distinct system of their own in the "core" of the Kuroshio and its extension. But I think it more likely that they are opportunistic species in the sense that Hutchinson (1967, pp. 364-365) has used that term, except that they "respond" to something other than weather. In addition to the brevity of discussion of Warm Water Cosmopolites, I have said little of the ecology of the blank areas of Figure 14. This is because I do not believe them to be true ecosystems but rather ecotones in which horizontal "immigration" of allochthonous individuals and nutrients and "emigration" of autochthonous individuals and nutrients are happening on a grand scale. The biogeographic evidence from eastern and western boundary currents indicates that this is so, as does the presence of faunal mixtures along the zonal boundaries. Further support for this comes from the studies of physical water mass mixing.

Thus, if ecosystems are thought of as being systems in which interactions and feedback loops between the essential components are the primary regulators of the state of the system, then ecotones are systems of a very different sort. This is not to say that interactions do not occur in them, perhaps even in a quasi-systematic way, but the actual state of the system and changes in state must depend much on large-scale, horizontal, advective processes. This means that these "systems" are very much open systems and probably can be understood only if the horizontal advection of nutrients, biomass, and species is understood, plus the *in situ* interactions and loops. This complicates the problem

Table 1. A comparison of some of the characteristics of five of the biotic provinces of the North Pacific

Biotic province	Percent spp endemic	Diversity, (no. spp)	Equitability	Variability zooplankton standing crop	Primary productivity (gC/m²/yr)	Productivity peak	Degree stratified	Range to top of thermocline (meters)	Leakage	Type of phytoplankton limitation	Neritic influence
Subarctic	High	Low	Low	High >2x	>75 <80	Spring	Low, but deep halocline	20 to >130	Very high	Light	Great
Transition	Moderate	Moderate	Moderate	Low <2x	>40 <75 ??	Late spring to early summer	Low, no halocline	20 to 120	Very high	Probably N	Very little
Central	Low	High	High	Low	≅40 ?	Winter	High, but halocline shallow in summer, deep in winter	30 to 110	Very low	N and P	None
Eastern Tropical Pacific	Low	High	Moderate	Moderate 2x	75	Two peaks, spring and fall	Complex	10 to 60 but complex	Moderate	N	Moderate to great
Equatorial	Moderate	High	Moderate	Low	>100 <220 ?	None ?	Both high and very low	35 to 65 and 60 to 140	Moderate	?	Very little
Warm Water Cosmopolites	....	High	High	Areally great, temporally ?	....	Complex	Complex	Complex	Low to moderate	....	Complex

The data for this table came from many sources, some of them quantitative and objective and others qualitative and subjective. I have indicated the most dubious of these characteristics with a question mark (?). However, some of the others are only slightly less dubious and all are in great need of further quantification.

Diversity is in terms of length of species list.

Equitability refers to the "evenness" of the rank order of zooplankton species abundance and may or may not be true of the phytoplankton and nekton.

Variability of zooplankton standing crop refers to temporal variability on a seasonal or longer basis.

Primary productivity per year has been measured over an entire year only in the Subarctic and Eastern Tropical Pacific. The other values are estimates based on short term (a few days or weeks) measurements and extrapolated over the entire year. These extrapolated values were weighted by factors derived from what we know of the seasonality of chlorophyll standing crops.

Degree stratified refers to the "strength" of the thermocline in terms of the steepness of the temperature gradient. It is related to the degree of vertical mixing of deeper, nutrient-rich water.

Range to top of thermocline is a measure of the thickness of the mixed layer, but south of 40° N the thermocline "topography" is much more complicated than indicated here.

Leakage refers to the proportion of the standing crop of plankton and nutrients estimated to be carried out of the system of its origin per unit time.

Type of phytoplankton limitation refers to the fundamental factors regulating turnover rates and specifically excludes grazing. N means nitrate limitation; P means phosphate limitation.

Neritic influence includes coastal upwelling as in the Eastern Tropical Pacific, coastal runoff as in the Subarctic, and island effects.

Although not included in this table because of our relative lack of information, I would expect the South Pacific Central, Transition, and Subantarctic provinces to be similar to those of the north in most (but not all) characteristics. The Warm Water Cosmopolites (WWC) have been included for comparative purposes.

enormously, for we know little of water mass mixing rates and even less of the mixing rates of such nonconservative properties as nutrients and populations. Along with this practical difficulty is the theoretical one that species which have evolved in one of the core systems are exposed to very different sets of biotic and physical conditions when transported out of that system into one of the ecotones. We can expect that their roles would be quantitatively *and* qualitatively different. For example, the main population of *Euphausia pacifica* is in the Subarctic ecosystem, but a sizable fraction of this population is swept down into the California Current along with Subarctic water. This euphausiid reproduces while in the California Current, but as its populations are moved to the south they decrease in abundance and the species eventually disappears. One can reasonably assume that the average speed of drift of a population of *E. pacifica* from the Subarctic gyre to a point off Baja, California, where it disappears, is 0.25 knots. This represents a transit time of about 200 days or enough time for, at most, only two generations (Brinton, pers. comm.). During the course of this drift, the *E. pacifica* populations are exposed to a continuously shifting complex of associated species and a changing physical regime. This is hardly enough time for evolutionary adaptation to the very different ecological conditions within this current. What is true of *E. pacifica* is also more or less so for *most* of the other species of Subarctic macrozooplankton and perhaps micronekton. Most of them have much larger gene pools well outside of this ecotone where the "rules of the game" are quite different.

As I pointed out earlier, there are some species of both zooplankton and nekton which are endemic to this area. In the case of zooplankton these tend to live quite near shore and do not represent a significant proportion of the biomass of the California Current in general. These and the endemic nekton are probably also opportunistic species. One might expect that these, like all opportunists, would be experiencing alternating periods of great success and relative failure. Evidence for this comes from the study of Soutar and Isaacs (1969), who show paleontological evidence that both the sardine, *Sardinops caerula,* and the anchovy, *Engraulus mordax,* two California Current endemics, have had nonrandom, noncyclical, oscillations equal to a factor of 10 or more in population size over the past 1800 years. In addition, they show that *Limacina helicina*, a Subarctic pteropod with a distribution almost identical with that of *Euphausia pacifica,* has varied by a factor of about 100 over the same time period. *L. helicina's* populations seem to be partly regulated by input from the north as are those of *E. pacifica.* If this is so, then there must be strong oscillations in this input. Independent evidence for this comes from Wickett (1966), who has shown that the large variations in the annual concentrations of zooplankton off southern California varied directly with the calculated southerly components of Ekman transport in the previous year at 50° N, 140° W. Many of the species in this southern portion of the current are not of Subarctic origin, rather they are Warm Water Cosmopolites or are from the Central province and are probably responding to the input of Subarctic-derived dissolved nutrients via the phytoplankton. Farther north, however, a large fraction of the biomass of California Current zooplankton is of Subarctic or Transition affinity. Other independent evidence of variations of the input of water (and its contained nutrients and plankton) comes from the studies of Saur (1972), who has examined the 65-year record of anomalies of sea level differences, Honolulu-minus San Francisco. He points out that "the currents are geostrophically related to the sea level difference so that a positive current index indicates above normal currents, i.e., stronger flow to the south around the eastern limb of the anticyclonic gyre of the North Pacific Ocean." There have been large, nonseasonal, positive and negative anomalies. Saur has identified seven "climatic periods" in this time series. These are "characterized by the variability and the mean level of the current index." They vary in duration from 96 to 180 months. In addition, there are episodes of even greater anomaly (both positive and negative) within these climatic periods ranging from 1 to 11 months' duration. Thus, not only does the input of water, nutrients, and fauna to this ecotone happen on a massive scale, it is highly variable with time. We do not, as yet, have this kind of information for the other ecotones of the Pacific. However, there is a very close coupling of the California Current to the rest of the circulation, and it seems unlikely that the magnitude of mixing or its variability is any less in the other boundary zones than it is here.

It would seem, then, that if we are to understand the workings of these ecotones in terms of what regulates the variations of abundance of species and biomass, we will need to know a great deal more about the quantitative nature of horizontal advection as well as *in situ* processes.

Since we do not, at the present time, understand the processes and events in even the semiclosed ecosystems of the gyres and recirculation systems, it is unlikely that we will be able to achieve this understanding for the "wide open" ecotones very readily. For here we must not only determine the factors regulating *in situ* changes, but simultaneously determine the role of large-scale advection of water, nutrients, and populations and, further, if we wish a predictive capability, the *causes* of this variation in advection. This seems to be a problem in global meteorology rather than ecology.

Literature Cited

Arrhenius, G. 1963. Pelagic sediments. In: *The Sea,* Vol. 3, pp. 655-727, M. N. Hill, ed. Interscience Publ., New York.

Brinton, E. 1962. The distribution of Pacific euphausiids. Bull. Scripps Inst. Oceanogr., *8(2)*: 51-270.

Cushing, D. H. 1971. Upwelling and the production of fish. In: *Advances in Marine Biology,* Vol. 9, pp. 255-326, F. S. Russell and Sir Maurice Young, eds. Academic Press, New York.

Dice, L. R. 1952. *Natural Communities.* University of Michigan Press, Ann Arbor.

Dodimead, A. J., F. Favorite, and T. Hirano. 1963. Review of the oceanography of the Subarctic Pacific region. In: *Salmon of the North Pacific Ocean, Part II.* International North Pacific Fisheries Comm. Bull. 13, pp. 1-195.

Fager, E. W., and J. A. McGowan. 1963. Zooplankton species groups in the North Pacific., Science, *140*(3566): 453-460.

Hutchinson, G. E. 1967. *A Treatise on Limnology.* Vol. II: *Introduction to Lake Biology and the Limnoplankton.* John Wiley & Sons, Inc., New York.

King, J. E. 1954. Variations in zooplankton abundance in the Central Equatorial Pacific, 1950-1952. Sympos. on Marine and Freshwater plankton in the Indo Pacific, pp. 1-8. Diocesan Press, Madras.

McGowan, J. A. 1971. Oceanic Biogeography of the Pacific. In *The Micropaleontology of Oceans,* pp. 3-74, B. M. Funnell and W. R. Riedel, eds. Cambridge Univ. Press, Cambridge.

Miller, C. B. 1970. Some environmental consequences of vertical migration in marine zooplankton. Limnol. and Oceanogr., *15*(5): 727-741.

Parsons, T. R., and R. J. Le Brasseur. 1968. A discussion of critical indices of primary and secondary production for large scale ocean surveys. Calif. Coop. Oceanic Fish. Invest. Reports XII, pp. 54-63.

Riedel, W. R., and B. M. Funnell. 1964. Tertiary sediment cores and microfossils from the Pacific Ocean floor. Quart. J. of the Geolog. Soc. London, *120:* 305-368.

Saur, J. F. T. 1972. Monthly sea level differences between the Hawaiian Islands and the California coast. Fishery Bull., *70*(3): 619-636.

Soutar, A., and J. D. Isaacs. 1969. History of fish populations inferred from fish scales in anaerobic sediments of California. Calif. Coop Oceanic. Fish. Invest. Reports XIII, pp. 63-70.

Tully, J. P., and F. G. Barber. 1960. An estuarine analogy in the sub-Arctic Pacific Ocean. J. Fish. Res. Bd. Canada, *17:* 91-122.

Venrick, E. L., J. A. McGowan, and A. W. Mantyla. 1973. Deep maxima of photosynthetic chlorophyll in the Pacific Ocean. Fishery Bull., *71:* 1.

Walsh, J. J. 1972. Implications of a systems approach to oceanography. Science, *176*(4038): 969-975.

Wickett, W. P. 1967. Ekman transport and zooplankton concentrations in the North Pacific Ocean. J. Fish. Res. Bd. Canada, *24* (3); pp. 581-594.

Wyrtki, K. 1966. Oceanography of the Eastern Equatorial Pacific Ocean. In: H. Barnes, ed., *Oceanogr. Mar. Biol. Ann. Rev.,* 1966, *4*, pp. 33-68.

The Control of Ecosystem Processes in the Sea

Timothy R. Parsons and Bodo R. de Lange Boom
Institute of Oceanography
University of British Columbia
Vancouver, B.C., Canada

A GREAT DEAL has been written about various factors which control ecosystem processes in the sea. In attempting to discuss the subject in a single article we shall limit ourselves to pelagic communities and assume that other marine communities (including benthic communities, coral reefs, salt marshes, and littoral communities) are subjects requiring special treatment.

While the pelagic environment itself contains many communities, the mechanisms contributing to the control of these communities are essentially similar in nature but different in effect. Thus the controlling mechanisms, which collectively produce an ecosystem, essentially compete with each other in deciding which biological processes shall be allowed to occur in any environment; as Dunbar (1960) has noted, this is a rather different concept than the accepted competition between individuals or specific populations as determinate factors in terrestrial ecology. Thus the changeability of the marine environment may cause wide fluctuations in animal populations of the sea, geographic displacement of productive zones, and variations in the food chain at the planktonic level affecting the survival of higher organisms (Uda, 1961; Shelbourne, 1957). As a general appraisal we may assume, as Dunbar (1960) suggested, that all marine ecosystems may be subject to irregular oscillations and that small oscillations are generally bad for the system, while large oscillations may be lethal. An ecosystem can then be characterized by the number and type of control mechanisms, such as predators, spatial patchiness, and nutrient regeneration, which tend to

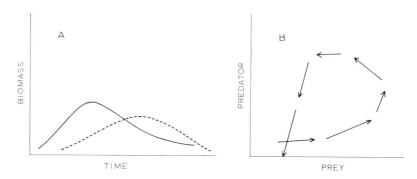

Figure 1. An ecosystem consisting of a single prey (e.g., a phytoplankter ——————) and a single predator (e.g., a zooplankter - - - - - - -). Changes in biomass with time (A) and predator/prey interaction (B) (from Maly, 1969).

dampen out oscillations by offering various ways in which the living components of the ecosystem can escape from overindulging in their own specialities.

In the simplest case of a single predator/prey association, the predator in pursuit of a prey may follow the course of events depicted in Figure 1 and cause its own demise by exhausting its food supply (Maly, 1969). Mechanisms which contribute to the stabilization of natural systems have been discussed extensively by terrestrial ecologists (e.g., Rosenzweig and MacArthur, 1963). The problem is to identify the mechanisms which exert similar controls on aquatic ecosystems.

Factors Controlling Ecosystem Processes in the Sea

The differentiation between a "factor controlling" an ecosystem process and an "ecosystem process" is often a matter of semantics. Thus the amount of phytoplankton available may control the growth of zooplankton; zooplankton themselves then exert a control on the phytoplankton. The roles of control and the resulting process are reversed. In some cases, however, there is no feedback, or if there is a feedback mechanism, it is operative through one or more additional identifiable components of the system. This is apparent in the effect of vertical stability of the water column as a controlling factor on phytoplankton production. Phytoplankton production does not itself influence vertical stability, although remotely a large phytoplankton crop may absorb more light and increase the temperature stratification of the surface layers.

	1. Nutrients	2. Vertical stability	3. Advection	4. Radiation	5. Extinction coefficient	6. Turbulence & diffusion	7. Temperature	8. Salinity	9. Nutrient regeneration	10. Phytoplankton production	11. Zooplankton production	12. Fish production	13. Food chain length	14. Transfer efficiencies	15. Food chain stability
	PHYSICAL/CHEMICAL EFFECTS ON THE PHYSICAL/CHEMICAL ENVIRONMENT								BIOLOGICAL EFFECTS ON THE PHYSICAL/CHEMICAL ENVIRONMENT						
1. Nutrients	X	+	+	O	O	+	O	O	O	+	O	O	O	O	O
2. Vertical stability	O	X	+	O	O	+	+	+	O	O	O	O	O	O	O
3. Advection	O	+	X	O	O	+	+	+	O	O	O	O	O	O	O
4. Radiation	O	O	O	X	+	O	+	O	O	O	O	O	O	O	O
5. Extinction coefficient	O	+	+	O	X	+	O	O	O	+	O	O	O	O	O
6. Turbulence & diffusion	O	+	+	O	O	X	O	O	O	O	O	O	O	O	O
7. Temperature	O	+	+	+	+	+	X	O	O	O	O	O	O	O	O
8. Salinity	O	+	+	O	O	+	+	X	O	O	O	O	O	O	O
	PHYSICAL/CHEMICAL EFFECTS ON THE BIOLOGICAL ENVIRONMENT								BIOLOGICAL EFFECTS ON THE BIOLOGICAL ENVIRONMENT						
9. Nutrient regeneration	O	O	O	O	O	O	+	+	X	+	+	+	O	O	O
10. Phytoplankton production	+	+	+	+	+	+	+	+	+	X	+	O	O	O	O
11. Zooplankton production	O	O	O	O	O	+	+	+	O	+	X	+	+	+	+
12. Fish production	O	O	O	O	O	O	+	+	O	O	+	X	+	+	+
13. Food chain length	O	O	+	O	O	O	O	O	O	+	+	+	X	+	+
14. Transfer efficiencies	O	O	O	O	O	O	O	O	O	+	+	+	+	X	+
15. Food chain stability	O	O	O	O	O	+	O	O	+	+	+	+	+	+	X

Figure 2. The interaction of biological and physical/chemical factors in the control of aquatic ecosystems (+ represents an interaction).

In Figure 2 an attempt has been made to identify fifteen components of a marine pelagic ecosystem which may interact both as controlling factors and processes, and in so doing they may define the type of ecosystem that is allowed to persist in any one environment. While the figure represents a continuous spectrum of components, it is also apparent that it can be divided broadly into physical/chemical (or largely abiotic factors) and biological factors; in doing this the figure is subdivided into four regions of influence, namely the physical/chemical effects on the physical/chemical environment; the biological effects on the physical/chemical environment; the physical/chemical effects on the biological environment; and the biological effects on the biological environment. Interaction between the same components has been blanked out; this may not be correct in all cases but generally the feedback of a component on itself is accomplished through some secondary mechanisms. The fifteen components identified in Figure 2 are not necessarily a complete list, but the combination is sufficiently large to form the basis for a discussion. In addition, some of the components are readily identified (e.g., temperature) while others are difficult properties to determine with a comparable degree of accuracy (e.g., stability of the food chain). A discussion of the four regions of interaction within the system is contained in the following sections.

I. Physical/Chemical Effects on the Physical/Chemical Environment

The control of ecosystem processes is heavily dependent on the movement of water masses. Physicists have described the properties of water movement using various terms which can usually be well defined in principle but may be difficult to apply in practice due to the complexity of the physical environment. Terms used to describe the physics of water movement include advection, diffusion, turbulence, and stability; these, in turn, are both governed by, or may affect, other physical properties of the environment, such as radiation in the sea, temperature, salinity, and the attenuation of light. Since these processes are important factors in the control of ecosystems in the sea, some elementary explanation of their properties may be useful.

Advection (affects 1,2,5,6,7,8)

The term *advection* needs clarification. Some authors use the word convection to describe the same physical process. In its more limited use convection refers to the circulation caused by variations in fluid density. The heated air which rises above a hot stove is an example of this process. Advection is the process by which properties of a fluid (such as temperature or nutrient concentration) are carried past a fixed point in space due to the motion of the fluid. Fluid motion may be induced by wind, tide, river flow, or pressure gradients.

Advection appears as one of the terms in the differential equations expressing conservation laws. For some quantity, Θ, in an incompressible fluid, the conservation equation may be written as,

$$\frac{\partial \Theta}{\partial t} + \mathbf{u} \cdot \nabla \Theta = F \qquad \text{I-1}$$

where t is time, $\mathbf{u} = (u, v, w)$ is the fluid velocity, and F is a source function including both sources and sinks. In Cartesian coordinates

$$\mathbf{u} \cdot \nabla \Theta \equiv u \frac{\partial \Theta}{\partial x} + v \frac{\partial \Theta}{\partial y} + w \frac{\partial \Theta}{\partial z} \qquad \text{I-2}$$

In general Θ may be a scalar or vector quantity specified in terms of position $\mathbf{x}$ and time t. Equation I-1 states that the time rate of change of Θ at a point $\mathbf{x}$ plus the advective rate of change is equal to the net rate of production of Θ.

Advection affects turbulence in two ways. First, the mean flow carries the turbulence along with it. Secondly, if there is a shear in the flow, then this shear may transfer energy to the turbulence. As an illustration we can consider an estuary with brackish river water flowing out over the saline deep water of the estuary. If we ignore other sources of turbulence, such as waves, then the turbulence at a fixed point in the upper layer consists of the turbulence carried to the point by the river water plus the turbulence produced by the velocity shear.

Advection can also affect the vertical stability of a column of water by introducing water of a different density. This density difference may be due to differences in salinity or temperature or both. If the density gradient is such that the low density water lies above the more dense water, then the water column will be stable with the stability increasing as the magnitude of the density gradient increases. Conversely, if the density gradient is of opposite sign, the water column will be unstable with the instability increasing as the magnitude of the gradient increases. Again, a good example can be found in an estuarine system. The surface water in fjords is less saline than the deep water. Thus this upper layer has a lower density than the deep water, producing a stable system. The stability is shown by the fact that an up or down inlet wind changes the thickness of the upper layer but leaves the layer itself intact.

Salinity, temperature, nutrient concentration, and extinction coefficients are all affected by advection. Nutrients may be brought up from deeper water when upwelling takes place. Similarly, the temperature and salinity may be changed by upwelling. The extinction coefficient will be affected by the intrusion of turbid water in coastal areas.

Diffusion and turbulence (6 on 1,2,3,5,7,8)

Diffusion may be described as the tendency for a nonuniform distribution of a property to reach equilibrium. In more precise terms, if Θ is some property of a fluid, and there exists a gradient $\nabla \Theta$, then diffusion acts to make the gradient go to zero. In its restricted use, diffusion usually refers to molecular diffusion.

Hinze (1959) defines turbulence as follows: "Turbulent fluid motion is an irregular condition of flow in which the various quantities show a random variation with time and space coordinates, so that statistically distinct average values can be discerned." Turbulent motion derives its energy from external forcing functions and acts in a manner similar to molecular diffusion. At the macroscopic scale, the turbulent diffusion dominates over the molecular diffusion, unless the level of turbulence is extremely low.

Including molecular diffusion and turbulence, equation I-1 becomes

$$\frac{\partial \bar{\Theta}}{\partial t} + \bar{u} \cdot \nabla \bar{\Theta} = \nabla \cdot (K \nabla \bar{\Theta} - \overline{u' \Theta'}) + F \qquad \text{I-3}$$

where $\bar{\Theta}$ and $\bar{u}$ are the average values, Θ' and u' are the turbulent fluctuations, K is the molecular diffusion coefficient, other symbols are as defined previously, and the overscore denotes an average.

The number of variables may be reduced by introducing an eddy or turbulent diffusion coefficient, A, and writing

$$-\overline{u' \Theta'} = A \nabla \bar{\Theta} \qquad \text{I-4}$$

However, one must be very cautious about the use of turbulent diffusion coefficients since generally they are functions of time, space, and conditions of the turbulent medium. Observations indicate that the coefficient increases with increasing scale size.

Turbulence affects the advection by removing energy from the mean flow. This energy goes into the turbulent fluctuations. Similarly, the diffusion of momentum (viscosity) removes energy from the flow. The result of this is a decrease in the velocity of the fluid.

Turbulent and molecular diffusion both affect the stability of a body of water through changing the density structure by the transport of salt and heat. It turns out that the coefficient of thermal conductivity (diffusion) is about 1×10^{-3} gm cal/sec cm °C, while the coefficient of diffusion for salt is approximately 2×10^{-5} gm/sec cm (Sverdrup et al., 1942). Thus heat diffuses about 100 times more rapidly than salt. Under the right conditions, this can result in instabilities in an otherwise stable situation. The turbulent diffusion coefficients are generally much larger than the molecular diffusion coefficients; the vertical turbulent diffusion coefficients may have values of up to 10^2 gm/sec cm, while in the horizontal they may reach 10^8 gm/sec cm for oceanic scales (Sverdrup et al., 1942). For smaller phenomena lesser values are appropriate.

The net effect of turbulent and molecular diffusion is to lead to a homogeneous distribution of properties in the water. This holds for temperature, salinity, nutrients, and extinction coefficients.

Vertical stability (2 on 1,3,5,6,7,8)

Stability in a physical system refers to the response of the system to perturbations. A system is said to be stable if, upon being perturbed, the system remains in the neighbourhood of its initial state. That is, it may oscillate about its initial state as with a pendulum, yet still be stable. Hesselberg (1918) defined the stability, E, of a mass of water as

$$E = \frac{1}{\rho} \frac{\delta \rho}{dz} \qquad \text{I-5}$$

where ρ is the density at depth z, (z positive downward), $\delta \rho$ is change in density of the surrounding water when a water volume is moved a distance dz. Thus $E > 0$ signifies stable, $E = 0$, neutral, and $E < 0$ unstable stratification. In actual fact the situation is not so simple since heat conduction, friction, and diffusion were not considered in the above equation. Neumann (1948) indicates that in the sea, neutral stability actually occurs for slightly negative values of E.

The stability of a water column may change through changes in the density structure caused by the advection of water of a different density, temperature changes, or changes in salinity. Stability can affect the advection by either aiding or inhibiting the flow. For example, upwelling of nutrients is weakened if the water column is very stable since the amount of energy required to raise the deep water is increased by the high stability. Similarly, vertical turbulent diffusion is inhibited by a stable stratification. An unstable stratification does not remain long since any slight disturbance will cause the system to seek a stable configuration.

Since a strong stratification inhibits turbulent transport, such properties as the temperature, salinity, nutrient concentration, and extinction coefficient will change much more slowly than in the case of a weak or neutral stratification. For example, the appearance of a shallow summer thermocline in mid-latitudes produces a stable stratification which prevents the heat from being mixed to depth in the absence of storms.

36 THE BIOLOGY OF THE OCEANIC PACIFIC

Extinction coefficient (5 on 4 and 7)

The extinction coefficient, $k(\lambda)$, is defined by the relation

$$dQ(\lambda)/dz = k(\lambda)Q(\lambda) \qquad \text{I-6}$$

where $dQ(\lambda)/dz$ is the rate of change of light intensity, $Q(\lambda)$, with depth and λ is the wavelength of the light. The value of $k(\lambda)$ varies with the wavelength of the light. $k(\lambda)$ is large for infra-red and ultra-violet light which makes up about 50 percent of the sun's energy spectrum at the sea surface on a cloudless day. Consequently, the extinction coefficient of light passing the first few centimeters can be averaged from $k(\lambda)$ values for light between 400 and 700 nm.

It is possible to write $k(\lambda)$ in terms of various components

$$k(\lambda) = (k' + k_b + k_p + k_d)(\lambda) \qquad \text{I-7}$$

where k' is the extinction of pure sea water, k_b is that due to living things (e.g., phytoplankton), k_p is that due to particulate material (e.g., silt), and k_d is that due to dissolved chemicals (pigments).

The extinction coefficient determines how rapidly light is attenuated in the water. In clear oceanic water about 16 percent of the total incident energy remains at a depth of 10 m, while at the same depth in turbid coastal water less than 1 percent may remain. Thus the availability of energy for biological processes depends upon the local extinction coefficient and how it varies with time.

Since the extinction coefficient is a measure of the energy absorbed in the water, it plays a part in determining the temperature of a volume of water. In weakly stratified systems the mixing of the water will dominate over the absorption, but in strongly stratified systems found in coastal areas the extinction coefficient determines the depth of heating due to incoming radiation.

Radiation (4 on 7)

The net radiation energy input to the sea, Q_N, can be written as,

$$Q_N = Q_d + Q_s - Q_b - Q_r \qquad \text{I-8}$$

where Q_d is the direct solar radiation at the sea surface, Q_s is the sky radiation at the sea surface, Q_b is the effective back radiation from the sea, and Q_r is the radiation reflected from the sea surface.

Q_d depends on the altitude of the sun and the cloud cover. At low altitudes the radiation is decreased due to absorption in the atmosphere and the fact that the radiation is spread over a larger horizontal area. Clouds reflect some of the radiation back into space and also scatter the incident radiation. Q_s, the scattered solar radiation, also depends on the sun's altitude, although not as strongly as Q_d. With a clear sky and a high sun, Q_d is about 85 percent and Q_s is 15 percent of the incident radiation. This changes to $Q_d = 60$ percent and $Q_s = 40$ percent for the sun 10 degrees above the horizon. On a dark, rainy day the incident radiation can be reduced to less than 10 percent of that on a clear day.

The effective back radiation, Q_b, is the difference between the "black body" radiation of the sea surface and the long wave radiation from the atmosphere. It appears that the sea surface radiates nearly like a black body, following Stefan's law (i.e., the radiated energy is proportional to the fourth power of the absolute temperature of the sea surface). For the sea most of this energy lies in the infra-red band. The long wave radiation from the atmosphere depends mainly on the water vapour content of the atmosphere. Thus Q_b depends mainly on the sea surface temperature and the relative humidity near the sea surface. Q_b decreases both with increasing temperature and increasing relative humidity as well as with increasing cloud cover. Most of the long wave radiation is absorbed within the first few millimeters of the sea.

The amount of reflected solar radiation depends upon the sun's altitude, while the reflected sky radiation does not. At an altitude of 90° the total reflected radiation is about 3 percent, while at 5° it is about 40 percent. The above holds for a smooth sea; in the presence of waves the reflection is increased somewhat for a low sun.

The radiation that is absorbed in the sea goes mainly into heating the water near the surface. Even in the clearest ocean water most of the heating occurs in the first meter. For an energy input to the water of 1,000 gm cal/cm², there is a temperature increase of 6.2° C for the clearest and 7.7° C for turbid water in the first meter. An energy input of 1,000 gm cal/cm² corresponds to the mean solar energy input over a three-day period in the tropics.

Some of the radiation energy is also changed to chemical energy through photosynthesis. Marine plants use the energy in the blue to yellow range of the spectrum (450 nm to 600 nm) for photosynthesis. Light in this range has the greatest penetrating power in the sea.

Temperature and salinity (7 and 8 on 2 and 3; 7 on 4 and 8)

The main processes by which the oceans gain or lose heat are radiation, atmospheric convection, and the condensation and evaporation of water. Geothermal heating, chemical heating, and the transformation of kinetic energy to heat are negligible.

Temperature mainly acts to change the density of the water. Thus thermal convection may be started and the stability of the water column may be affected. Temperature can affect the salinity by causing water to evaporate at the sea surface, thereby increasing the salinity. Occasionally water condenses into the sea from the air, decreasing the salinity.

Salinity is considered to be a conservative property although this is not strictly true. The salt content of the seas is slowly increasing. Variations in salinity occur due to precipitation, river runoff, and evaporation. As with temperature, salinity changes cause changes in the water density. Thus vertical stability is affected and convection currents can be set up.

II. Physical/Chemical Effects on the Biological Environment

Vertical stability, extinction coefficient, and radiation on phytoplankton production (2, 4, and 5 on 10)

Riley (1946) and Sverdrup (1953) have discussed the combined effect of vertical stability, extinction coefficient, and radiation on phytoplankton production. Both authors essentially use the same formulation to derive the average light intensity in the water column and to compare this with the compensation light intensity. Thus if water is mixed vertically and homogeneously to a depth D, and the average extinction coefficient for photosynthetic light in the water column is k, then the incoming photosynthetic radiation ($0.5\ Q_N$) is attenuated to give an average light intensity $\bar{Q}_N$ in the mixed layer, such that

$$\bar{Q}_N = \frac{0.5\ Q_N}{kD}\ (1 - e^{-kD}) \qquad\qquad \text{II-1}$$

Several difficulties are apparent in the application of this formula. One of these is to decide on a value for the surface photosynthetic radiation. In the above expression this has been given as 0.5 of Q_N, which is the total radiation. Thus approximately half of the total radiation is considered to be in the region 400 to 700 nm on a bright sunny day (Strickland, 1958). However, on a cloudy day a much larger fraction of the total radiation is in the photosynthetic region. Ideally this value ($0.5\ Q_N$) should be measured directly with a pyranometer masked with a filter to remove infra-red and ultra-violet radiation. Alternatively some adjustment has to be made to allow for cloudy days. Also, at low sun angles ($< 15°$) or under stormy sea conditions an appreciable amount of radiation is reflected from the sea surface. The extinction coefficient, k, is an average value for the water column; the value is large for light at the red end of the spectrum and with the exception of turbid coastal waters, the value generally employed in oceanic studies is for blue/green light (about 430 nm).

The depth of mixing may be determined in a variety of ways, such as the depth to the bottom of the principal thermocline or halocline, or from the maximum value obtained from the expression ($\partial\sigma/\partial z \times 10^3$), which represents the change in density with depth. While it is recognized that these terms may not give the same depths, the actual choice may be decided by what data are available and how clearly they show the vertical stratification of the water column.

The 24-hour compensation light intensity, Q_c, for phytoplankton is generally in the range 0.003 to 0.009 ly/min (Hobson, 1966; Parsons et al., 1969). This value is obviously very critical to the application of the formula and more information is required on compensation light intensities. If the value of $\overline{Q}_N$ is above the compensation light intensity, then the average net productivity for the water column can be determined from a photosynthesis (P) versus light intensity (Q) curve (e.g., Ryther, 1956). The P versus Q curve can be represented by an equation such as:

$$P = P_{max} \ (1 - \exp \ [\beta \ (Q_c - Q)] \ ) \qquad\qquad \text{II-2}$$

where P is the photosynthesis at light intensity, Q, P_{max} is the maximum photosynthesis per unit chlorophyll, Q_c is the compensation light intensity, and β is a constant characteristic of the population.

The use of equation II-2 raises another problem, however, and this is to determine the response constant (β) and the P_{max} values for the phytoplankton population in the area of study. The slope at the compensation intensity (Q_c) is proportional to β since $\triangle P/\triangle Q$ at Q_c equals $\beta \ P_{max}$; this slope can be shown to be dependent on the species of phytoplankton present. The slope is generally large for chlorophyceae, smaller for diatoms, and least for dinoflagellates, the range of values being about 0.05 to 1.0 mgC(mg Chl-a hr klux)$^{-1}$. P_{max} values (mgC/mgChl a/hr) are dependent on environmental conditions, such as nutrients and temperature, and can range from less than 1 to about 6 (although higher values are also found).

Finally, having established the general form of the P versus Q curve and the light intensity with depth, the production for the water column can be determined from a knowledge of the depth distribution of the phytoplankton (chlorophyll a), and the multiple of standing stock of chlorophyll and chlorophyll specific productivity can be integrated graphically to give the productivity per m^2/unit time.

Temperature and salinity effects on biological processes (7 and 8 on 9,10,11,12)

Temperature and salinity have long been recognized as factors which exert a controlling influence over plant and animal populations in the sea. At a very early stage in the development of biological oceanography, the words eurythermal, stenothermal, euryhaline, and stenohaline were used to describe whether the tolerance of organisms to changes in temperature and salt content was wide or narrow. Bacteria, which are the primary agents in nutrient regeneration in the sea, are often referred to as being halophilic or halophobic according to whether they will grow on media containing more or less than 2 percent NaCl, respectively (Larsen, 1962). Temperature tolerance of commercial fish species has been summarized by Uda (1961); for 21 species of commercial fish Uda shows that seven species inhabit a temperature range of about 2 to 7° C, nine inhabit a range of about 12 to 18° C, and five species are generally found in water $>20°$ C. The combination of temperature and salinity as determinate factors in the distribution of plankton has been discussed by Bary (1959 and 1963). The author summarized temperature, salinity, and plankton data in figures (T-S-P diagrams) which showed that plankton were related to certain water bodies rather than to any geographical location.

Temperature also affects the rate of biological processes. The simplest expression for the effect of temperature is a two or threefold increase in the rate of a biological reaction for a 10° C rise in temperature (over the range of temperature tolerance for the process in question). This is an oversimplification, however, and a discussion of the response of biological processes to temperature has been given by McLaren (1963, 1965, and 1966). From his studies it was concluded that the best representation of original data was generally obtained by Bélehrádek's empirical formula

$$V = a \ (t + \alpha)^b \qquad\qquad \text{II-3}$$

where V is the rate of a biological process, t is temperature and a, b, and α are constants. These constants have been discussed by McLaren, who considered that α represented a "biological zero" expressing the temperature characteristic of the process when $V = 0$, b represented the degree of curvature and reflected the general dependence of all biological processes on temperature, and a represented the units in which V was measured, as well as being an expression of any surface to volume restriction governing the rate of exchange of a rate-limiting metabolite across a surface membrane.

Effects of advection, turbulence, and diffusion on biological production (3 on 10 and 13; 6 on 10 and 11)

The advective effects of the major currents of the oceans are primarily responsible for producing large-scale differences in plankton productivity throughout the hydrosphere. The general mechanism causing zones of high production involves the transport of deep (about 200-300 m) nutrient-rich water to the surface layers where there is sufficient light for primary production. Such large-scale upwelling of water is caused either directly or indirectly by the wind—directly by causing a divergence of the surface water, such as may occur near coastlines, indirectly by producing surface currents which interact to produce divergences. The direction of the wind relative to the coastline or the direction of the currents relative to each other determines whether a divergence (upwelling) or a convergence (downwelling) will be formed (due to Coriolis force, see Pickard, 1964). Areas of divergence are particularly important and examples may be found in the Pacific Ocean off the coast of Peru and along the equator. Smaller scale patchiness in plankton distributions can be observed due to smaller upwelling processes. These are apparent in connection with land masses where eddy currents may be produced in the lee of islands, around headlands, or as a result of changes in the bottom topography (LaFond and LaFond, 1971). Wind-induced plankton patchiness may be partly accounted for in terms of Langmuir circulation. Stavn (1971) has offered an explanation for several types of plankton distributions which can be observed under the effect of weak and strong winds. These distributions include the passive arrangement of nonmotile particles in upwelling helices if the particles are heavier than water, and in downwelling helices for particles lighter than water. When the phototactic response of zooplankton is included, however, these organisms may be found to accumulate either in weak downwelling helices where they swim upwards, or in strong upwelling helices where they try to swim down away from the high surface radiation (negative phototaxis). Zooplankton may also accumulate at the bottom of helices at intermediate water velocities.

In addition to water movement, other components causing plankton patchiness are the differences in the reproduction rate of the plankters, strong physical gradients such as haloclines and thermoclines, social behaviour of species, and coactive factors between species including predation patterns.

Dispersions can be expressed statistically (Cassie, 1962) as

$$\hat{c} = (s^2\text{-}m)/m^2 \qquad\qquad \text{II-4}$$

where s^2 is the variance of the plankton distribution and m is the mean concentration of plankton; $\hat{c}$ can be regarded as an index of dispersion. From the point of view of food chains and feeding relationships, some patchiness component in plankton distributions is important in decreasing the mean distance a predator has to travel in collecting (plankton) prey items. This is especially important in oligotrophic waters where the mean prey abundance, expressed as some average concentration over distance, is often too low to support the food density required by plankton feeders. Thus patchiness is an important quantitative component in feeding relationships (e.g., equation III-11).

III.   Biological Effects on the Biological Environment

Factors affecting food chain stability (6 and 9 to 14 on 15; 15 on 11, 12, 13, and 14)

Stability is an important property of any food chain, although unfortunately it is not readily defined and even more difficult to measure. It is a property which is determined by the environment and by the physiological reaction of the fauna and flora to the environment; if the total effect of these factors is towards stability, then stability of the ecosystem is in a sense a result of a feedback process which assures the continuation of the various trophic levels of production (i.e., 15 on 11, 12, and 13). Hurd et al. (1971) defined stability as the ability of a system to return to its ground state after a small external perturbation. The subject of stability has been discussed recently by Dickie and Mann (in press), who used Weiss' (1969) definition: a system exhibits stability when the variance of a property of the total system is less than the variance of the components. This definition has some merit in that it can be used as an index such that

$$S = \frac{N\,V}{(v_1 + v_2 + \ldots + v_i)} \qquad \text{III-1}$$

where S is an index of stability of a system property having a variance V, and a number of components (N) making up the system with variances of $v_1, v_2, \ldots v_i$. An advantage to the use of the above ratio as an index of stability is that it compares the relative value of the input and output to a system without having to express the absolute output amplitude of the oscillations. Thus one system which shows regular oscillations over time of several hundred percent may be as stable as a system output which shows small oscillations with time. Patten (1961 and 1962) considered the stability of a plankton community and its environmental parameters. The stability index evolved by Patten (1962) was expressed as

$$S' = \frac{\sum\limits_{j=1}^{m} \det P_j}{\sum\limits_{j=1}^{m} (s/\bar{x})_j} \qquad \text{III-2}$$

Stability will occur for $S' > \varepsilon$ where the value of ε depends upon the system under consideration ($\varepsilon > 0$). ε is a value of S' representing the boundary value between stability and instability.

Now

$$P_j = \begin{bmatrix} P_{id} & P_{ii} \\ P_{dd} & P_{di} \end{bmatrix} \qquad \text{III-3}$$

which is the matrix of transition probabilities for the jth of m variables, P_{id} being the probability for a decrease in the value of the variable following an increase, P_{ii} that for an increase following an increase, and so on (see Patten, 1961) where the P's are determined from observation ($P_{id} + P_{ii} = P_{dd} + P_{di} = 1$).

Also $\det P_j$ is calculated from

$$\det P_j = P_{id} P_{di} - P_{ii} P_{dd} \qquad \text{III-4}$$

and $(s/\bar{x})_j$ is the standard deviation divided by the mean for the jth variable.

In equation III-2 the denominator indicates the probability of a variable oscillating about some mean value (that is, $\det P_j$ will be positive when an increase of a variable is followed by a decrease, and vice versa, with $-1 \leq \det P_j \leq 1$).

Secondly, the expression $s/\bar{x}$ gives a measure of the variation of a component or the size of the oscillations (stability, S' being larger for smaller standard deviations about the mean). Equation III-2 is obviously a more comprehensive representation of stability than equation III-1 and values for S' have been determined by Patten (1961 and 1962). In comparing S' for components of the physical environment and plankton community, Patten found that the plankton (S'_p) was 6.7 times more stable than the environment (S'_e). This is not an unreasonable quantitative expression since the plankton must to some extent

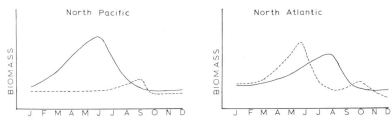

Figure 3.   Phytoplankton (- - - - -) and zooplankton (————) changes in biomass with time in the North Pacific and North Atlantic oceans (adapted from
Heinrich, 1962).

absorb the "shock" of environmental change and if $S'_p < S'_e$ the
plankton community would collapse.

Factors which contribute to stability in a food chain relationship
have been discussed by several authors (Rosenzweig and MacArthur,
1963; McAllister et al., 1972). These include feeding migrations, prey
refuges, resting stages, inhomogeneous distributions, abiotic factors,
and diversity of the food web. Thus in a plankton community the nocturnal migration of zooplankton into the surface layers is a stabilizing
influence on the type of predator/prey association shown in Figure 3.
In this case the nonuniform nocturnal grazing by zooplankton allows
for the growth of phytoplankton during the day, an effect which assures a maximum food supply (Petipa and Makarova, 1969). In other
examples, the overwintering of a population of *Calanus plumchrus* at
several hundred meters in the Pacific Ocean stabilizes the life cycle of
this animal; patchiness of food organisms increases stability by allowing for greater feeding in the patches while the concentration of food
organisms outside the patches is allowed to increase above a feeding
threshold level (p_o, Eq. III-12); temperature as an abiotic factor may
limit a predator's feeding and growth ability in a situation such as
Figure 1; the number of components in a food web will alter the diversity of prey available to an animal so that the exhaustion of one
prey item is prevented by the presence of another.

Diversity (which can be measured separately by several methods)
is therefore a major component contributing to stability. In the interaction figure, the stabilizing factors mentioned above are inherent characteristics of the component identified (i.e., 6 to 14). Turbulence (6),
for example, can either induce or destroy various forms of patchy
plankton distributions including those found under conditions of Langmuir circulation (Stavn, 1971); nutrient regeneration (9) is a feedback mechanism which allows for a continual supply of phytoplankton
even after the nutrient concentration of the water has apparently become exhausted (e.g., in tropical environments); the variations in

phytoplankton (10), zooplankton (11), and fish (12) production, such as the size of organisms and selectivity of predators for prey items (Parsons and Lebrasseur, 1970), will influence, together with the overall length of the food chain (13), the stability of the system. Transfer efficiencies (14), in the sense of the nutritional adequacy of a prey, may also cause wide differences in growth and survival of organisms (Provasoli et al., 1959).

Conversely, factors contributing to instability will be the absence of some of the processes described above. However, one major component of instability which is not apparent in Figure 2 is the need for synchronization of certain events in nature. For example, the ability of barnacle larvae to survive in the Firth of Clyde was shown by Barnes (1956) to be dependent on the time of arrival of a bloom of *Skeletonema costatum;* mortality in the absence of this bloom was close to 100 percent, which resulted in the failure of larvae to settle out as a new barnacle population.

From these observations it is quite apparent that variability in components contributes to the stability of a biological system which is the opposite process to that contributing to the stability of a machine (Weiss, 1969) or a highly controlled ecosystem, such as is found in aquaculture projects. In controlled operations, such as machines, the variance of each component adds up to describe the variance in the performance of the total unit. Only by imposing maximum control over the components is it possible to predict the operation of the machine. In contrast, in biological systems the components vary widely but the system itself (if stable) may show only small (or if large, regular) oscillations. This contrast is brought out when attempts are made to stimulate a system using a computer model which inadequately describes the buffering action of the natural environment. Thus small changes in a component of such models can be shown to produce practically any change in a population, however unrealistic such changes may be in a real ecosystem (McAllister, 1970). In contrast, it appears that a natural aquatic ecosystem is made up of highly variable components which interlock at some point in their oscillating states to allow for a flow of energy up the food chain; the exact position of the interlocking point on any mechanism is maintained by fluctuations about a mean, rather than by rigid control of an absolute value. One corollary of these observations is that a perturbation applied to the top of a food chain should have more effect on the nature of the food chain than a perturbation of similar magnitude applied to the bottom of the food chain. This may be seen in part by two large-scale perturbations applied to sockeye-producing lakes in British Columbia. In the first experiment (Foerster and Ricker, 1941; Foerster, 1968) 90 percent of the sockeye predators were removed from a lake, resulting in about a 300 percent increase in the biomass of sockeye produced in the lake. In a second experiment (Parsons et al., 1972; Barraclough and Robinson, 1972) nutrients in a large sockeye-producing lake were increased by 100 percent; this resulted in only about a 30 percent increase in the biomass of sockeye, due to depensatory components (particularly a temperature effect). While these two experiments are only very crudely comparable, they serve to illustrate the large effect of a perturbation applied to the top, in contrast to the bottom, of an aquatic food chain.

Transfer efficiencies, production, and food chain length (10 to 15 on 14, and 14 on 11 to 15)

Transfer efficiencies between different trophic levels in the food chain are difficult to determine *in situ* and estimates are generally based on experimental data. For physiological studies, growth efficiency of an animal is expressed as

$$K_2 = \triangle W / \triangle t \, AR \qquad \text{III-5}$$

where $\triangle W / \triangle t$ is the increase in an animal's weight per unit time, R is the animal's ration (weight/time), and A is the assimilation efficiency (Winberg, 1956). The latter term, A, is defined as

$$A = \frac{\text{amount of food ingested}}{\text{amount of food eaten}} \qquad \text{III-6}$$

The assimilation efficiency of different food materials has been summarized by Sushchenya (1970) for algal-zooplankton relationships and by Winberg (1956) for zooplankton-fish relationships with the result that a value of 0.8 with a range from 0.4 to 0.95 may be generally accepted for both steps in the food chain.

The value of K_2 from experimental data has been shown to vary with changes in both the type of ration and the quantity eaten. A general relationship between K_2 and R has been given for fish feeding by Palaheimo and Dickie (1966) as

$$ln \, K_2 = -a -bR \qquad \text{III-7}$$

There is also general confirmation for this expression given by Sushchenya (1970) for zooplankton feeding. The terms a and b in the above expression are in part related to the type of diet so that different growth efficiencies can be expected depending on the type of food organism present in any environment at the time of study. K_2 also decreases with the age of organisms so that the highest efficiencies are

Table 1. Potential fish production at different ecological efficiencies
and trophic levels

Primary production (mg C/m²/yr)	Trophic levels	Efficiency (%)	Fish production (mg C/m²/yr)
100	2	20	4
100	5	10	0.001

observed in young, actively growing organisms and efficiencies approaching zero are found in mature adults.

Unfortunately, K_2 is not directly representative of *in situ* conditions because environmental variability due to abiotic factors, together with the mixture of prey items in any environment, generally decreases an animal's growth efficiency. Thus growth efficiencies of 30 to 40 percent which may be observed in the laboratory are not necessarily representative of field conditions. For field studies, efficiency between trophic levels is defined as the ecological efficiency (E), such that

$$E = \frac{\text{amount of food extracted from a trophic level}}{\text{amount of food supplied to a trophic level}} \qquad \text{III-8}$$

Slobodkin (1961) suggested that ecological efficiencies were generally about 10 percent, and this figure has been used widely. However, it is probable that, at least at the lower trophic levels involving the plankton community where predator and prey are in close contact over extended periods of time, ecological efficiencies may be in the range 10 to 20 percent.

Schaeffer (1965) and Ryther (1969) have considered the effect of changes in ecological efficiencies on the productivity at various trophic levels. An adaptation of Schaeffer's results is shown in Table 1.

In the first line of Table 1 a food chain having two trophic levels above the phytoplankton (e.g., phytoplankton → copepods → herring) at 20 percent ecological efficiency is compared with a food chain having 10 percent ecological efficiency and five trophic levels [e.g., phytoplankton → herbivorous zooplankton (copepods) → carnivorous zooplankton (euphausiids) → planktivores (myctophids) → piscivores (squid) → salmon]. The resulting potential fish production is very obviously different in these two food chains. This difference can be further accentuated by having different levels of primary production which may vary by a factor of about 6 between stable water columns in the open ocean (about 50 g C/m²/yr) and some upwelling areas and coastal zones (about 300 g C/m²/yr).

In conclusion, it is apparent that the type of organisms in an environment may largely determine food chain efficiencies (e.g., 10, 11, 12, and 13 reactions on 14, Figure 2) while the resulting efficiencies themselves affect the overall production of organisms (e.g., 14 on 11, 12, and 13). In addition, both the food chain length and the stability of an ecosystem (through the diversity of organisms making up a food chain) are intimately connected with the transfer efficiencies.

Effects of nutrients on phytoplankton production (1 and 9 on 10)

Certain nutrients in the sea are present in abundant quantities and their availability to phytoplankton does not generally cause any growth restriction, except under a few unusual circumstances. Biologically important elements and compounds included in this category are Na, Mg, K, Ca, Cl, $SO_4^=$, H_2O, and CO_2. Many trace elements may also be present in abundant supplies, but this subject has not been extensively researched; elements which might be included in this list are Mn, Cu, Zn, B, V, Co, and Mo. One obvious exception among these elements appears to be iron, and sufficient information now exists (Menzel and Ryther, 1961; Tranter and Newell, 1963) to indicate that the availability of iron may sometimes limit phytoplankton growth, especially in subtropical and tropical seas. In addition, certain trace organic compounds, particularly vitamins, such as vitamin B_{12}, thiamine, and biotin, may determine the type of phytoplankton which grow in certain environments, although such compounds are not required for truly autotrophic growth. Other trace growth factors which have yet to be identified also affect phytoplankton growth. Thus Johnston (1963 and 1964) found that phytoplankton growth was inhibited by presence or absence of a "factor" in certain types of Atlantic water and similar effects have been found in recently upwelled water off Peru (Barber et al., 1971) and in water from 100 m in the tropical Atlantic (Smayda, 1971).

In general, however, major nutrients affecting phytoplankton growth are nitrogen, phosphorus, and silicon. Silicon is an essential element for diatoms, but it cannot be regarded as essential to the phytoplankton community where flagellates and other phytoplankton can grow in its absence. Phosphorus is an essential nutrient for all phytoplankton, but phosphorus regeneration is rapid (Antia et al., 1963) and thus its turnover rate usually assures an adequate supply of this element. Nitrogen, on the other hand, is regenerated much more slowly than phosphorus in the absence of zooplankton (Antia et al., 1963); in the presence of zooplankton, nitrogen is regenerated through the feeding activity of zooplankton. The nitrogen excretory product of zooplankton is primarily ammonia, which may be utilized directly by some phytoplankton or converted back to nitrite and nitrate by bacteria.

From some of the earliest studies on phytoplankton physiology (Ketchum, 1939; Spencer, 1954), it was shown that the uptake of nutrients was concentration dependent and that the total biomass density of phytoplankton produced was proportional to the total concentration of a growth-limiting nutrient. Since nitrogen can most often be shown to be the rate-limiting nutrient, the kinetics of nutrient uptake have been centered on the uptake of various forms of this element (i.e., ammonia, nitrite, and nitrate).

Caperon (1967) and Dugdale (1967) showed that nutrient uptake could be defined in terms of Michaelis-Menton kinetics, such that

$$V = V_{max} \, S/(K_s + S) \qquad \text{III-9}$$

where V and V_{max} are the rates of nutrient uptake and the maximum rate of nutrient uptake, respectively; S is the concentration of nutrient and K_s is the Michaelis constant equal to the nutrient concentration at $V_{max}/2$. K_s is an important property of a phytoplankton species which may be related both to the environmental conditions (MacIsaac and Dugdale, 1969) and the size of the organism (Eppley and Thomas, 1969). Thus high values of K_s have been found for large diatoms in coastal waters and low values of K_s have been found for small flagellates in oceanic waters.

The equation given above may be related directly to the phytoplankton growth constant (Thomas, 1970), providing the amount of nutrient per cell is constant. This is generally true of phytoplankton during their exponential phase of growth, but for the uptake of vitamin B_{12}, Droop (1970) showed that growth was proportional to the concentration of vitamin B_{12} in the cell and not to the concentration of vitamin in the medium. Thus some caution may be necessary in applying Michaelis-Menton kinetics as a description of nutrient uptake under all conditions.

One further aspect of nutrient uptake by phytoplankton is to consider whether the nutrient uptake of one species could affect the rate of nutrient uptake of another. Hulburt (1970) considered that each algal cell could be visualized as the center of a volume of water in which nutrient depletion decreases from the cell outwards; thus nutrient zones would have to overlap for one species (with its characteristic K_s) to affect the uptake of another species. Hulburt (1970) estimated that the zones of nutrient uptake would overlap at cell concentrations $>3 \times 10^8$ cells/litre. Since most natural populations do not exceed 10^6 cell/litre, the likelihood that one species could affect the nutrient uptake of another is generally small.

Interaction of zooplankton with phytoplankton and fish production (11 on 9, 10, 12, and 13; 10 and 12 on 11)

Zooplankton occupy a central position in the marine food chain and as such they can directly influence both primary and tertiary producers.

Nutrient regeneration by zooplankton is an important function in maintaining phytoplankton growth during periods of high stability in temperate waters, or throughout the year in most tropical waters. Several authors (e.g., Marshall and Orr, 1961; Pomeroy et al., 1963) have studied the release of phosphate and organic phosphorus by zooplankton; depending on the size of the zooplankton, excretion rates are probably between 10 and 100 percent of the animal's body phosphorus per day. In the presence of very small zooplankton (e.g., Protozoa) the rate of phosphorus regeneration may be several orders of magnitude greater (Johannes, 1965). Nitrogen appears to be generally excreted as ammonia (Harris, 1959; Corner and Newell, 1967) or as organic nitrogen compounds (Johannes and Webb, 1965; Webb and Johannes, 1969). Ctenophores and jellyfish, which appear to be a terminal level of production in the marine environment, may also serve a useful purpose in nutrient regeneration by recycling the microcrustacean members of the zooplankton community (Fraser, 1962).

The transfer of phytoplankton to zooplankton and of zooplankton to fish in the marine food chain appears to be concentration dependent and may be generally described by Ivlev's (1961) empirical equation

$$r = R(1\text{-}e^{-kp}) \qquad\qquad \text{III-10}$$

where r is the ration consumed by an animal per unit time, R is the maximum ration which an animal can consume per unit time, p is the density of prey organisms, and k is a constant. Although Ivlev (1961) developed this relationship for fish feeding, it is also apparent that the relationship can be used for zooplankton-phytoplankton feeding (Parsons et al., 1967; Conover, 1968; Sushchenya, 1970). Two modifications to this relationship which may apply in certain instances are that patchiness of prey may increase its availability to a predator and, secondly, that populations of prey items may not be grazed down to zero prey density (as implied by Ivlev's equation) but to some prey density (p_o) at which feeding ceases. Experimental evidence for the first of

these modifications was given by Ivlev (1961), who included patchiness of prey in his equation as

$$r = R \ (1-\exp \ [-k'p-k''s]\ ) \qquad \text{III-11}$$

where k' and k'' are constants and S is a measure of patchiness. The
second modification appears in the experimental data of some investigators (Parsons et al., 1967; Adams and Steele, 1966; Nassogne,
1970) but not in the data of others (e.g., Paffenhöfer, 1970) ; in this
case it may be desirable to introduce a threshold prey concentration at
which feeding starts (p_o) such that

$$r = R(1-\exp \ [k(p_o-p)\ ]\ ) \qquad \text{III-12}$$

In addition to the effect of prey density on animal feeding, there
are other factors, such as selectivity of an organism for a prey item,
which must also be measured quantitatively. Selectivity may depend on
a number of factors, such as the speed of the prey, the presence of
defensive coverings (such as spines), the size of the prey organism and
its palatability. These factors may be brought together in some general
expression of selectivity, such as was used by Ivlev (1961) :

$$E = (r_i - p_i)/(r_i + p_i) \qquad \text{III-13}$$

where r_i is the fraction of the prey item in the ration and p_i is the
fraction of the prey item in the total population of the prey.

More recently the feeding of aquatic organisms has been considered over longer time periods in which it is necessary to consider an
animal's overall metabolic requirements. As an example of these considerations, Kerr (1971) derived a formula for the feeding of fishes in
which he considered growth efficiency, metabolic and hunting requirements of the predator, the internal cost of food utilization, and prey
factors such as size, concentration, and swimming speed.

IV.   Biological Effects on the Physical/Chemical Environment

Chlorophyll Effect on the Extinction Coefficient (10 on 5) and Phytoplankton Production on Nutrients (10 on 1)

An empirical relationship between the concentration of chlorophyll
a and the extinction coefficient (k) was determined by Riley (1956) as

$$k = 0.04 + 0.0088C + 0.054C^{2/3} \qquad \text{IV-1}$$

where C is the chlorophyll a concentration in mg/m^3. This appears to
be a valid relationship for shallow waters where other factors which
affect k are small (e.g., silt and soluble organic material). Instead of

using the above formula, a specific relationship can be derived as a polynomial curve for a particular environment.

The effect of phytoplankton on nutrients under stable conditions generally leads to reduction in the concentration of one nutrient to a level below which the growth of phytoplankton is essentially zero. However, in practice, nutrient regeneration maintains a small standing stock of phytoplankton even in the most oligotrophic waters. The relationship between nutrient concentration and phytoplankton growth has been discussed in Section III.

Discussion

On the basis of the previous discussion it is apparent that there are many components which feature in the control of marine ecosystems; the effect of some of these, such as light, are fairly well known, while others, such as nutrient regeneration, are only beginning to receive quantitative expression. However, throughout the hydrosphere certain areas serve as good examples of cases where one or two controlling mechanisms tend to predominate for part or all of the year.

In the northeast subarctic Pacific Ocean it is generally recognized that zooplankton grazing exerts a controlling influence on phytoplankton standing stock during the summer months. In the oceanic waters of the Gulf of Alaska one can find abundant nitrate and phosphate concentrations in midsummer, but a low standing stock of phytoplankton due to zooplankton grazing. Quite a different situation is apparent in the North Atlantic, where the standing stock of phytoplankton shows a marked increase in the spring causing a decrease in the availability of nutrients (Figure 3; McAllister et al., 1960; Heinrich, 1962). In another example, nutrient regeneration as a controlling factor is particularly important in maintaining phytoplankton productivity in a highly stratified water column. Since nutrient regeneration by bacteria is highly temperature dependent (Sorokin, 1971), the effect is most apparent in near-shore temperate waters during the summer (e.g., Steemann Nielsen, 1958), and in coastal subtropical and tropical waters during most of the year (Steemann Nielsen and Jensen, 1957).

Light and the extinction coefficient of water are undoubtedly the two most predominant factors in regulating the *type* of productivity which may occur in different environments. This relationship for various oceanic and coastal areas has been discussed by Steeman Nielsen and Jensen (1957) and Ryther (1962). Ryther has developed an earlier idea (Riley, 1941) that the total productivities of different water columns (oceanic, coastal, and estuarine) do not in fact differ greatly per m². Thus the compensatory mechanisms between light attenuation due to particulate and soluble material (other than phytoplankton—see Eq. I-7), nutrient exhaustion, and the total amount of light available tend to depress differences in the levels of total production. For example, in spite of differences of several orders of magnitude in the primary productivity per m³ of tropical and temperate waters, the actual total difference in production between a coastal temperate ecosystem and an oceanic tropical ecosystem is probably no greater than a factor of about 4 (Ryther, 1962). A much greater difference in the two ecosystems lies in the food web which develops as a result of the different physical/chemical environments. Thus the crucial point in ecosystem control in the sea comes back to a complex interaction between organisms. Some attempt to understand these interactions has been made through sensitivity analysis, which allows one to determine analytically how a given variable or parameter in a system will be affected by changes in some other variable or parameter (Patten, 1969).

Most of the current models available for sensitivity analysis deal with the marine ecosystem as a production system and largely exclude factors which contribute towards stability. It appears, in fact, that the greatest challenge to the biological oceanographer today is to study the ecological components which have led to the evolution and maintenance of aquatic ecosystems as stable parts of our total environment. The principal destabilizing effects which man is imposing on the marine ecosystem are removal of predators through commercial fishing and the potential pollution of the sea by pesticides, heavy metals, and plastics. On the one hand, fisheries scientists have been overly concerned with obtaining the "maximum sustainable yield" of a fishery. Population theory is largely employed in obtaining such information, while the environment which is "sustaining the yield" is ignored. On the other hand, the success of industrialization has been measured in terms of "gross national product," with little concern for what eventually happens to the "product"; since many industrial products are ending up in the sea, it becomes relevant to ask whether this is resulting in a change in ecosystem control sufficient to cause a destabilization of the system. A purely speculative question might be asked as to whether there has been an increase in the ctenophore and jellyfish population due to the removal of more efficient plankton predators such as whales and herring. An understanding of processes leading to the control of marine ecosystems is therefore of paramount importance, not merely for our production-oriented economy, but more especially for assuring the survival of the sea as a stable environment.

Literature Cited

Adams, J. A., and J. H. Steele. 1966. Shipboard experiments on the feeding of *Calanus finmarchicus* (Gunnerus). In: *Some Contemporary Studies in Marine Sciences,* pp. 19-35, H. Barnes, ed., Allen and Unwin Ltd., London.

Antia, N. J., C. D. McAllister, T. R. Parsons, K. Stephens, and J. D. H. Strickland. 1963. Further measurements of primary production using a large-volume plastic sphere. Limnol. Oceanogr., *8:* 166-183.

Barber, R. T., R. C. Dugdale, J. J. MacIssaac, and R. L. Smith. 1971. Variations in phytoplankton growth associated with the source and conditioning of upwelling water. Inv. Pesg., *35:* 171-193.

Barnes, H. 1956. *Balanus balanoides* (L.) in the Firth of Clyde: The development and annual variation of the larval population, and the causative factors. J. Anim. Ecol., *25:* 72-84.

Barraclough, W. E., and D. Robinson. 1972. The fertilization of Great Central Lake. III. Effect on juvenile sockeye salmon. Fish. Bull., *70:* 37-48.

Bary, B. M. 1959. Species of zooplankton as a means of identifying different surface waters and demonstrating their movements and mixing. Pac. Sci., *13:* 14-34.

Bary, B. M. 1963. Distribution of Atlantic pelagic organisms in relation to surface water bodies. In: Marine Distributions, M. J. Dunbar, ed. Royal Soc. Canada, Sp. Publ. No. 5, pp. 51-67.

Caperon, J. 1967. Population growth in micro-organisms limited by food supply. Ecology, *48:* 715-722.

Cassie, R. M. 1962. Frequency distribution models in the ecology of plankton and other organisms. J. Anim. Ecol., *31:* 65-92.

Conover, R. J. 1968. Zooplankton—life in a nutritionally dilute environment. Am. Zoologist, *8:* 107-118.

Corner, E. D. S., and B. S. Newell. 1967. On the nutrition and metabolism of zooplankton. IV. The forms of nitrogen excreted by *Calanus.* J. Mar. Biol. Ass. U. K., *47:* 113-120.

Dickie, L. M., and K. H. Mann. 1972 (in press). Analysis of biological production systems.

Droop, M. R. 1970. Vitamin B_{12} and marine ecology. V. Continuous culture as an approach to nutrition kinetics. Helgoländer wiss. Meeresunters., *20:* 629-636.

Dugdale, R. C. 1967. Nutrient limitation in the sea: Dynamics, identification, and significance. Limnol. Oceanogr., *12:* 685-695.

Dunbar, M. J. 1960. The evolution of stability in marine environments. Natural selection at the level of the ecosystem. Amer. Natur., *94:* 129-136.

Eppley, R. W., and W. H. Thomas. 1969. Comparison of half-saturation constants for growth and nitrate uptake of marine phytoplankton. J. Phycol., *5:* 375-379.

Foerster, R. E., and W. E. Ricker. 1941. The effect of reduction of predaceous fish on survival of young sockeye salmon at Cultus Lake. J. Fish. Res. Bd. Canada, *5:* 315-336.

Foerster, R. E. 1968. The Sockeye Salmon, *Oncorhynchus nerka.* Bull. Fish. Res. Bd. Canada, 162.

Fraser, J. H. 1962. The role of ctenophores and salps in zooplankton production and standing crop. Rapp. Proces-Verbaux Reunions, Conseil Perm. Int. Explor. Mer., *153:* 121-123.

Harris, E. 1959. The nitrogen cycle in Long Island Sound. Bull. Bingham Oceanogr. Coll., *17:* 31-65.

Heinrich, A. K. 1962. The life histories of plankton animals and seasonal cycles of plankton communities in the oceans. J. Cons. Int. Explor. Mer., *27:* 15-24.

Hesselberg, T. 1918. Über die Stabilitätsverhältnisse bei vertikalen Verschiebungen in der Atmosphäre und im Meer. Ann. d. Hydr. u. Marit. Meteorol., *46:* 118-129.

Hinze, J. O. 1959. *Turbulence, An Introduction to its Mechanism and Theory.* McGraw-Hill, New York.

Hobson, L. A. 1966. Some influences of the Columbia River effluent on marine phytoplankton during January, 1961. Limnol. Oceanogr., *11:* 223-234.

Hulburt, E. M. 1970. Competition for nutrients by marine phytoplankton in oceanic, coastal, and estuarine regions. Ecology, *51:* 475-484.

Hurd, L. E., M. V. Mellinger, L. L. Wolfe, and S. J. McNaughton. 1971. Stability and diversity at three trophic levels in terrestrial successional ecosystems. Science, *173:* 1134-1136.

Ivlev, V. S. 1961. *Experimental Ecology of the Feeding of Fishes.* D. Scott, trans. Yale Univ. Press, New Haven.

Johannes, R. E. 1965. Influence of marine protozoa on nutrient regeneration. Limnol. Oceanogr., *10:* 434-442.

Johannes, R. E., and K. L. Webb. 1965. Release of dissolved amino acids by marine zooplankton. Science, *150:* 76-77.

Johnston, R. 1963. Seawater, the natural medium of phytoplankton. I. General features. J. Mar. Biol. Ass. U. K., *43:* 427-456.

Johnston, R. 1964. Seawater, the natural medium for phytoplankton. II. Trace metals and chelation, and general discussion. J. Mar. Biol. Ass. U. K., *44:* 104-116.

Kerr, S. R. 1971. A simulation model of lake trout growth. J. Fish. Res. Bd. Canada, *28:* 815-819.

Ketchum, B. 1939. The absorption of phosphate and nitrate by illuminated cultures of *Nitzschia closterium.* Am. J. Bot., *26:* 399-407.

LaFond, E. C., and K. G. LaFond. 1971. Oceanography and its relation to marine organic production. In: *Fertility of the Sea,* Vol. I, pp. 241-265, J. D. Costlow, ed. Gordon and Breach, New York.

Larsen, H. 1962. Halophilism. In: *The Bacteria. A Treatise on Structure and Function.* IV. The Physiology of Growth, pp. 297-342, I. C. Gunsalus and R. Y. Stanier, eds. Academic Press, New York.

MacIsaac, J. J., and R. C. Dugdale. 1969. The kinetics of nitrate and ammonia uptake by natural populations of marine phytoplankton. Deep-Sea Res., *16:* 45-57.

McAllister, C. D., T. R. Parsons, and J. D. H. Strickland. 1960. Primary productivity and fertility at Station 'P' in the northeast Pacific Ocean. J. Cons. Int. Explor. Mer, *25:* 240-259.

McAllister, C. D. 1970. Zooplankton rations, phytoplankton mortality and the estimation of marine production. In: *Marine Food Chains,* pp. 419-457, J. H. Steele, ed. Oliver and Boyd, Edinburgh.

McAllister, C. D., R. J. LeBrasseur, and T. R. Parsons. 1972. Stability of enriched aquatic ecosystems. Science, *175:* 562-564.

McLaren, I. A. 1963. Effects of temperature on growth of zooplankton and the adaptive value of vertical migration. J. Fish. Res. Bd. Canada, *20:* 685-727.

McLaren, I. A. 1965. Some relationships between temperature and egg size, body size, development rate and fecundity of the copepod, *Pseudocalanus*. Limnol. Oceanogr., *10:* 528-538.

McLaren, I. A. 1966. Predicting development rate of copepod eggs. Biol. Bull., *131:* 457-469.

Maly, E. J. 1969. A laboratory study of the interaction between the predatory rotifer *Asplanchia* and *Paramecium*. Ecology, *50:* 59-73.

Marshall, S. M., and A. P. Orr. 1961. On the biology of *Calanus finmarchicus*. XII. The phosphorus cycle: Excretion, egg production, autolysis. J. Mar. Biol. Ass. U. K., *41:* 463-488.

Menzel, D. W., and J. H. Ryther. 1961. Nutrients limiting the production of phytoplankton in the Sargasso Sea, with special reference to iron. Deep-Sea Res., *7:* 276-281.

Nassogne, A. 1970. Influence of food organisms on the development and culture of pelagic copepods. Helgoländer wiss. Meeresunters., *20:* 333-345.

Neumann, G. 1948. Über den Tangentialdruck des Windes und die Raugigkeit der Meeresoberfläche. Z. Meteorol., Vol. 2, No. 7/8 (Potsdam).

Paffenhöfer, G. A. 1970. Cultivation of *Calanus helgolandicus* under controlled conditions. Helgoländer wiss. Meeresunters, *20:* 346-359.

Palaheimo, J. E., and L. M. Dickie. 1966. Food and growth of fishes. III. Relations among food, body size and growth efficiency. J. Fish. Res. Bd. Canada, *23:* 1209-1248.

Parsons, T. R., R. J. LeBrasseur, and J. D. Fulton. 1967. Some observations on the dependence of zooplankton grazing and the cell size and concentration of phytoplankton blooms. J. Oceanogr. Soc. Japan, *23:* 10-17.

Parsons, T. R., K. Stephens, and R. J. LeBrasseur. 1969. Production studies in the Strait of Georgia. Part I. Primary production under the Fraser River plume, February to May, 1967. J. Exp. Mar. Biol. Ecol., *3:* 27-38.

Parsons, T. R., and R. J. LeBrasseur. 1970. The availability of food to different trophic levels in the marine food chain. In: *Marine Food Chains,* pp. 325-343, J. H. Steele, ed. Oliver and Boyd, Edinburgh.

Parsons, T. R., K. Stephens, and M. Takahashi. 1972. The fertilization of Great Central Lake. I. Effect of primary production. Fish. Bull., *70:* 13-23.

Patten, B. C. 1961. Preliminary method for estimating stability in plankton. Science, *134:* 1010-1011.

Patten, B. C. 1962. Improved method for estimating stability in plankton. Limnol. Oceanogr., *7:* 266-268.

Patten, B. C. 1969. Ecological systems analysis and fisheries science. Trans. Amer. Fish. Soc., *99:* 570-581.

Petipa, T. S., and N. P. Makarova. 1969. Dependence of phytoplankton production on rhythm and rate of elimination. Mar. Biol., *3:* 191-195.

Pickard, G. L. 1964. *Descriptive Physical Oceanography*. Pergamon Press, Oxford.

Pomeroy, L. R., H. M. Mathews, and H. S. Min. 1963. Excretion of phosphate and soluble organic phosphorus compounds by zooplankton. Limnol. Oceanogr., *8:* 50-55.

Provasoli, L., K. Shiraishi, and T. R. Lance. 1959. Nutritional idiosyncrasies of *Artemia* and *Tigriopus* in monoxenic culture. Ann. New York Acad. Sci., *77:* 250-261.

Riley, G. A. 1941. Plankton studies. III. Long Island Sound. Bull. Bingham Oceanogr. Coll., *7:* 1-93.

Riley, G. A. 1946. Factors controlling phytoplankton populations on Georges Bank, J. Mar. Res., *6:* 54-73.

Riley, G. A. 1956. Oceanography of Long Island Sound, 1952-1954. II. Physical oceanography. Bull. Bingham Oceanogr. Coll., *15:* 15-46.

Rosenzweig, M. L., and R. H. MacArthur. 1963. Graphical representation and stability conditions of predator-prey interactions. Amer. Natur., *97:* 209-223.

Ryther, J. H. 1956. Photosynthesis in the ocean as a function of light intensity. Limnol. Oceanogr., *1:* 61-70.

Ryther, J. H. 1963. Geographic variations in productivity. In: *The Sea,* Vol. 2, pp. 347-380, M. N. Hill, ed. Interscience Publ., New York.

Ryther, J. H. 1969. Photosynthesis and fish production in the sea. Science, *166:* 72-76.

Schaeffer, M. B. 1965. The potential harvest of the sea. Trans. Amer. Fish. Soc., *94:* 123-128.

Shelbourne, J. E. 1957. The feeding and conditions of plaice larvae in good and bad plankton patches. J. Mar. Biol. Ass. U. K., *36:* 539-552.

Slobodkin, L. B. 1961. *Growth and Regulation of Animal Populations* (Chap. 12). Holt, Rinehart, and Winston, New York.

Smayda, T. J. 1971. Further enrichment experiments using the marine centric diatom *Cyclotella nana* (clove 13-1) as an assay organism. In: *Fertility of the Sea,* Vol. 2, pp. 493-510, J. D. Costlow, ed. Gordon and Breach, New York.

Sorokin, J. I. 1971. On the role of bacteria in the productivity of tropical waters. Int. Rev. ges. Hydrobiol., *56:* 1-48.

Spencer, C. P. 1954. Studies on the culture of a marine diatom. J. Mar. Biol. Ass. U. K., *33:* 265-290.

Stavn, R. H. 1971. The horizontal-vertical distribution hypothesis: Langmuir circulations and *Daphnia* distributions. Limnol. Oceanogr., *16:* 453-466.

Steemann Nielsen, E., and E. A. Jensen. 1957. Primary oceanic production. The autotrophic production of organic matter in the oceans. Galathea Rep., *1:* 49-136.

Steeman Nielsen, E. 1958. A survey of recent Danish measurements of the organic productivity of the sea. Rapp. Cons. Explor. Mer, *144:* 92-95.

Strickland, J. D. H. 1958. Solar radiation penetrating the ocean. A review of requirements, data and methods of measurement, with particular reference to photosynthetic productivity. J. Fish. Res. Bd. Canada, *15:* 453-493.

Sushchenya, L. M. 1970. Food rations, metabolism and growth of crustaceans. In: *Marine Food Chains,* pp. 127-141, J. H. Steele, ed. Oliver and Boyd, Edinburgh.

Sverdrup, H. U., M. W. Johnson, and R. H. Fleming. 1942. *The Oceans, Their Physics, Chemistry and General Biology.* Prentice-Hall, Englewood Cliffs, N. J.

Sverdrup, H. U. 1953. On conditions for the vernal blooming of phytoplankton. J. Cons. Explor. Mer, *18:* 287-295.

Thomas, W. H. 1970. Effect of ammonium and nitrate concentration on chlorophyll increases in natural tropical Pacific phytoplankton populations. Limnol. Oceanogr., *15:* 386-394.

Tranter, D. J., and B. S. Newell. 1963. Enrichment experiments in the Indian Ocean. Deep-Sea Res., *10:* 1-9.

Uda, M. 1961. Fisheries oceanography in Japan, especially on the principles of fish distribution, dispersal and fluctuations. Calif. Coop. Oceanic Fish. Invest., *8:* 25-31.

Webb, K. L., and R. E. Johannes. 1969. Do marine crustaceans release dissolved amino acids? Comp. Biochem. Physiol., *29:* 875-878.

Weiss, P. 1969. The Living System: Determinism Stratified. In: *Beyond Reductionism,* pp. 3-55, A. Koestler and J. Smythies, eds. Hutchinson, London.

Winberg, G. G. 1956. Rate of metabolism and food requirements of fishes. Fish. Res. Bd. Canada Trans. Ser. No. 194.

Feeding Processes at Lower Trophic Levels in Pelagic Communities

B. W. Frost
Department of Oceanography
University of Washington
Seattle, Washington

THE ULTIMATE GOAL of biological oceanography is to understand and predict the distribution and abundance of organisms in the sea. Broadly speaking, this is a problem of population dynamics, and trophic interactions are of central importance among the factors affecting the rates of birth and death in populations. Marine pelagic ecosystems may be most sensitive to trophic interactions occurring at the lower levels of the pelagic food web, the herbivore and primary carnivore level (Steele, 1972). Trophic relationships at this level have been qualitatively known for some time (e.g., Hardy, 1924). However, because specific quantitative information is lacking about even the simplest pelagic food webs, the trophic-dynamic model (Engelmann, 1966) has been used to predict abundance of populations in higher trophic levels (Ryther, 1969; Gulland, 1970). These predictions can be improved only when details are available on major pathways for energy flow in pelagic ecosystems.

Even at the base of marine pelagic food webs, the species and size composition as well as abundance of phytoplankton must play an important role in trophic interactions and energy flow. Parsons et al. (1967) and Parsons and LeBrasseur (1970) show that not only are the herbivorous zooplankton influenced by how energy emanates in different pathways from the phytoplankton, but that the effects ramify into higher trophic levels as well.

Trophic interactions may have even more far-reaching consequences for pelagic communities than those which are directly evident in natality and mortality. The experimental manipulations of small freshwater ponds by Hall, Cooper, and Werner (1970) have clearly demonstrated that a single species of planktivorous fish may, through size-selective predatory activity, shape the structure of fairly complex natural communities of plankton. Brooks and Dodson (1965) conclude that in several lakes where another planktivorous fish, the alewife, has recently been landlocked, the species and size composition of the zooplankton have been drastically altered. From both of these studies it appears that competitive trophic interactions determine the species composition of the zooplankton when planktivorous fish are absent and usually a large-sized species dominates the zooplankton. Planktivorous fish preferentially remove the larger zooplankters, which presumably are the superior competitors, and this tends to promote greater species diversity in the zooplankton. Similar trends should be apparent in phytoplankton assemblages preyed upon by size-selective herbivores.

Size-selective predation may have other interesting results. For example, the size composition of food particles could be a prime determinant of individual and population growth. Kerr (1971b) attempts to demonstrate this in an analytical model of individual growth of fish; growth efficiency of fish increases markedly when the size of food particles increases without changing, or even decreasing, the abundance of available food. Apparently there is field evidence for this phenomenon (Kerr and Martin, 1970). Feeding preference, especially size-selective feeding, may dominate the feeding strategy of all planktonic consumers.

The preceding ideas form the basis of the following discussion of feeding processes at lower trophic levels in marine pelagic food webs. The emphasis will be on zooplankton, particularly particle grazers or filter feeders but also carnivores, and how individual and population growth are influenced by tactics and strategy of feeding. Generalizations which are made about the feeding behavior of particle grazers are likely to be applicable to certain invertebrate predators, e.g., copepods, so no distinction in feeding behavior is necessarily implied by referring to these consumers by different names. However, clear differences in feeding behavior are likely to separate visual and nonvisual predators. My remarks are intended to complement some recent reviews on feeding (Jorgensen, 1966; Conover, 1968; Monakov, 1972), but the aim of this paper is to indicate the dominant importance of size composition of prey to feeding processes and population dynamics of consumers. A portion of my original presentation is published (Frost, 1972); consequently, the treatment of some topics has been expanded and some new areas are explored.

Laboratory Culture of Zooplankton

Laboratory experimentation on zooplankton has only recently begun to produce results which may contribute to interpretation of field phenomena. Laboratory culture of important planktonic herbivores, both protozoan and metazoan, has been achieved only in the last decade. Techniques are still being perfected, judging from recent results on *Calanus pacificus* (= *helogolandicus*), for example (cf. Mullin and Brooks, 1970a, 1970b; Paffenhöfer, 1970). This sudden success is remarkable in view of the attention given to experimental studies of species of *Calanus* over the last half century. Two techniques appear to account for the breakthrough: (1) the use of large-volume containers with efficient stirring mechanisms, and (2) the use of large-sized species of phytoplankton as food particles. The effect of large container volumes is chiefly to reduce the rate of decrease of food concentration between additions of food. The possibility that large containers alleviate an effect of crowding of herbivores, as suggested by Paffenhöfer (1970), has never been conclusively demonstrated and seems unlikely in view of the results of Slobodkin (1954) and King (1967) on other types of particle grazers. Increasing the size of food particles, within allowable limits, ensures that the grazers can more easily obtain a daily ration. Batch culture techniques probably will continue to be the most efficient means of investigating feeding processes and population dynamics of the larger metazoan consumers, while continuous culture methods will be more useful for studies of protozoans (Hamilton and Preslan, 1970; Gold, 1971).

Laboratory culture of planktonic carnivores seems to require more specialized techniques, but recent results suggest that this is a promising area for research (Greve, 1972; Reeve and Walter, 1972).

Functional Response of Predator

Single prey species

The relationship between prey density and the number of prey consumed by a predator is called the functional response of the predator (see Holling, 1959). An approximately hyperbolic type of curve is the typical functional response for pelagic invertebrate particle grazers and predators and for small vertebrate predators (Ivlev, 1961; Reeve, 1964; Parsons et al., 1967, 1969; LeBrasseur et al., 1969; McAllister, 1970; Parsons and LeBrasseur, 1970). Three equations have been used to describe the functional response:

$$I = I_m (1 - e^{-\delta P}) \tag{1}$$

$$I = I_m P/(K + P) \tag{2}$$

$$I = aP/(1 + abP) \tag{3}$$

where I is the measured rate of consumption of prey at prey density P and I_m is the maximal rate of consumption of prey. Equation (1) was derived by Gause and applied by Ivlev (1961); δ is a proportionality constant specifying the rate of change of I with respect to P. Equation (2) is the Michaelis-Menton relationship commonly used to describe nutrient uptake rates of microorganisms; K is the prey concentration at which I equals $I_m/2$. Equation (3) is from Holling (1959); the constant a is the instantaneous rate of discovery of prey by the predator, and b specifies the time required to capture and eat a prey item.

All of these equations may provide acceptable approximations for computer models, but they appear to be of limited value for understanding the basis of the functional response. For example, equations (2) and (3) predict a maximal rate of prey consumption at an infinitely large prey density. This may be a valid representation of data from brief feeding experiments in which animals are starved prior to feeding trials (e.g., Corner et al., 1972; Frost, 1972), but when animals are continuously feeding they achieve a maximal rate of prey consumption at finite and comparatively low prey densities (e.g., Frost, 1972). Equation (1) provides a better fit to experimentally derived data, especially when few points are available (Parsons et al., 1967, 1969; LeBrasseur et al., 1969; McAllister, 1970; Parsons and LeBrasseur, 1970). However, for particle-grazing copepods the functional response may also be interpreted as two intersecting straight lines (Fig. 1), which implies that different mechanisms control consumption rates at different prey densities. Results (Frost, 1972) suggest that copepods filter at a somewhat variable, maximal rate when prey densities are below a critical density; consumption rate is limited by the filtering capacity of the animal. At prey densities above a critical level, con-

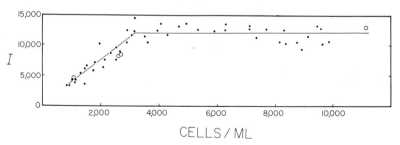

Figure 1. Functional response of adult females of *Calanus pacificus* feeding on the diatom *Thalassiosira fluviatilis*. I = cells eaten per copepod per hour. See Frost (1972) for details.

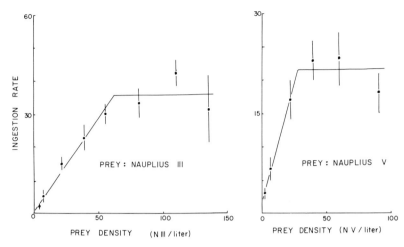

Figure 2. Functional response of adult females of the carnivorous calanoid copepod *Tortanus discandatus* feeding on nauplius stage III or nauplius stage V of *Calanus pacificus*. Unpublished data of J. W. Ambler.

sumption rate of copepods is determined by digestive processes, i.e., how fast the consumed material is processed and passes through the alimentary tract. As pointed out by Holling (1965), this rectilinear type of response curve may be typical of filter feeding crustaceans; it may also apply to small predatory calanoid copepods (Fig. 2). The response curve is characteristic of a nonvisual predator searching at random for prey and whose searching rate is not affected by prey density. This interpretation is consistent with direct observations of feeding behavior of freshwater cladocerans (McMahon and Rigler, 1963; Burns, 1968a). Equation (1) may better represent the functional response of visual predators, such as planktivorous fish (e.g., Ivlev, 1961), whose searching behavior is affected by prey density.

Assuming that the results of laboratory experiments are relevant to the field, claims of "excess" or "superfluous" feeding by natural populations of zooplankton (for a review, see Corner et al., 1972) do not seem consistent with the form of functional response described above. If superfluous feeding does occur, assimilation efficiency should decrease with increasing prey density (Beklemishev, 1962). Yet in laboratory cultures, continuously feeding grazers react to extraordinarily high densities of prey by adjusting their ingestion to a fixed maximal rate (Frost, 1972). In accord with this, assimilation efficiency and prey density are either independent or they are dependent in such a way that assimilation efficiency decreases with increasing prey density until a minimum value is reached at a finite prey density. A choice between these two alternatives cannot now be made from experimental results (Conover, 1966; Schindler, 1968; Butler et al., 1970; Sushchenya, 1970; Schindler, 1971; Corner et al., 1972).

Adams and Steele (1966) and Parsons et al. (1967) found that the feeding rate of particle-grazing copepods may be depressed at very low prey densities, but there are conflicting observations (McAllister, 1970; Paffenhöfer, 1971; Corner et al., 1972; Frost, 1972). Depression of feeding rates could depend on the size of food particles and might involve a feeding strategy in which energy gain per time spent feeding is optimized. Particle grazers might continue to feed at very low densities of large particles since large particles are efficiently captured, but they may cease filtering at relatively high densities of small particles. There are no observations to support this and, in fact, the apparent depression of feeding rates may be due purely to inadequate experimental methods for investigating feeding behavior at low food densities. The available evidence on directly measured filtering rates suggests that filter feeders, both benthic and pelagic, show maximal filtering rates when exposed to particle-free seawater or very dilute suspensions of food (Loosanoff and Engle, 1947; McMahon and Rigler, 1963; Davids, 1964; Burns, 1968a).

The functional response for a predator may vary with both the size of prey organisms and the size of the predator. The particle-grazing copepod *Calanus pacificus* feeds more effectively on large-sized diatoms and acquires its maximal daily ration at lower absolute food densities of larger diatoms (Fig. 3). The upper limit for the size of particles which a copepod can ingest is probably determined by body

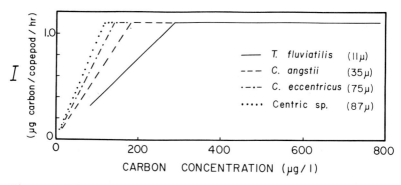

Figure 3. Effect of size (species) and concentration (as carbon) of food particles (diatoms) on ingestion rates (I) of adult females of *Calanus pacificus*. See Frost (1972) for details.

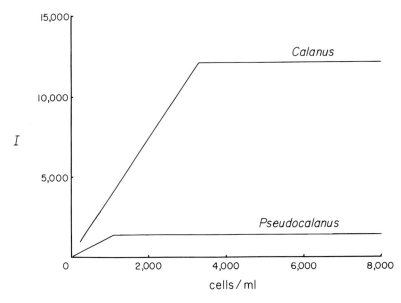

Figure 4. Functional response of adult females of *Calanus pacificus* and *Pseudocalanus* sp. feeding on the diatom *Thalassiosira fluviatilis*. *Calanus* weighs about 10 times as much as *Pseudocalanus* and attains its maximal ration at a higher density of cells than *Pseudocalanus*. I = cells eaten per copepod per hour. Results on *Calanus* from Frost (1972); feeding experiments with *Pseudocalanus* were carried out under identical conditions by P. M. Evans (unpublished).

size of the grazer, as in cladocerans (Burns, 1968b). The lower limit also may be physically set by the dimensions of filtering sieves. Thus a certain type of food particle may be large for an immature stage of a grazer and small for a mature stage of the same grazer. Functional response curves for different developmental stages of the grazer will show this effect primarily by variations in the prey density at which the maximal consumption rate is reached (see Fig. 2 in McMahon, 1965). Species of grazers of different size may show the same pattern (Fig. 4). In the case of predators it is not always clear that the effect of size of prey alone, and not prey species or susceptibility of prey to capture, is important. Nevertheless, Parsons and LeBrasseur (1970) found that juvenile salmon met their daily food requirements at relatively low densities of a large calanoid copepod but could not satisfy their nutritional needs even at excessively high densities of a small calanoid copepod. As the chaetognath *Sagitta hispida* grows, it feeds on progressively larger prey items (Reeve and Walter, 1972).

Multiple prey species

In the oceans, consumers at lower trophic levels are exposed to a wide spectrum of food particles; the spectrum typically has a shape grading from very abundant small particles to very rare large particles. Over broad oceanic areas the spectrum of particle sizes (between about 1μ and 100μ) may remain relatively constant geographically, vertically, and seasonally (Sheldon et al., 1972). If, indeed, the particle size spectrum is predictable, then there is opportunity for particle grazers and small predators to specialize on specific size ranges of food particles (Schoener, 1969). However, extension of Schoener's arguments, which involve consideration of energy gain per unit time spent feeding, to pelagic particle grazers leads to conflict with the assumptions embodied in the size-efficiency hypothesis of Brooks and Dodson (1965). Given some predictable size-frequency distribution of food particles, large and small species of grazers can coexist only at high food densities according to Schoener's theory; when food is scarce, small grazers would be favored in competition for food. In contrast, Brooks and Dodson (1965) suggest that large herbivores will outcompete small herbivores for food. They assume that particle grazers (specifically, cladocerans and copepods) of all sizes complete for fine particulate matter, but that larger grazers have a greater capacity for food collection and also can utilize larger particles than small grazers can. The basic differences between the two theories are shown schematically in Figure 5. For two different sizes (species) of grazers the outcome of interspecific competition for food hinges on which theory represented in Figure 5 is correct.

Observations on the feeding behavior of two species of calanoid copepods, *Pseudocalanus* sp. and *Calanus pacificus*, are relevant to Brooks and Dodson's size-efficiency hypothesis. *Pseudocalanus* is roughly an order of magnitude smaller in weight than *Calanus*, but it feeds in essentially the same way. The functional response has been determined for adult females of both species feeding on the diatom *Thalassiosira fluviatilis* (Fig. 4). *Pseudocalanus* achieves its maximal ration at about one-third the cell density required by *Calanus*. Further, at cell densities below those at which each species can obtain its maximal ration, the weight-specific filtering rate (ml swept clear/mg copepod dry weight/day) for *Pseudocalanus* is about two and one-half times greater than that for *Calanus*. Based on these two criteria, *Pseudocalanus* (the smaller grazer) feeds more effectively on *T. fluviatilis* than the larger *Calanus*. *T. fluviatilis* is clearly a small food particle for adult *Calanus* (Fig. 3).

Brooks and Dodson's evidence for the size-efficiency hypothesis was primarily drawn from studies on the feeding behavior of cladocer-

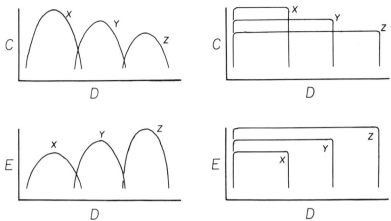

Figure 5. Two theories for the utilization of a spectrum of food particles by three species of grazers of different sizes (size of the grazers increases in the order X, Y, Z). *Upper graphs,* rate of food consumption per unit weight of consumer (C) versus size of food particle (D); food in all particle sizes is superabundant. Following Schoener's (1969) theory (left graph), grazers of different sizes feed most effectively at different positions on the particle-size spectrum; the shape of the curves is not critical but they may actually be skewed to the left and may decline more steeply to the right. In Brooks and Dodson's (1965) theory (right graph) grazers of all sizes feed on small particles, but large grazers also utilize food particles which are too large for small grazers; it is implicitly assumed that each grazer is an equally effective feeder on all sizes of particles which it can handle. *Lower graphs,* net energy gained per time spent feeding (E) versus size of food particles; food in all particle sizes is taken to be scarce, i.e., the density of particles of each size does not permit grazers to obtain their maximal ration. To include effects of foraging activity, the biomass of food particles is the same in all size classes of particles. In the left graph each curve may actually be skewed to the left and descend more sharply on the right.

ans. The hypothesis seems intuitively wrong or at least very simplistic; dominance in competitive interactions depends on many factors and is often unpredictable. Appropriate data for Figure 5 cannot be found in the literature. Frank's (1957) experiments on competition between two species of *Daphnia* provide one direct test which does not support the hypothesis—the smaller *Daphnia* always outcompeted the much larger species. A more recent test of the theory (Sprules, 1972) is inconclusive, and perhaps this is what we should expect. Further, Burns (1969) indicates that *Daphnia* may be more selective in its feeding behavior than previously was thought.

Calanoid copepods appear to feed selectively when presented with a choice of several food organisms. This behavior is evident when copepods are fed either mixed laboratory cultures of phytoplankton or natural phytoplankton (see Frost, 1972). Selective feeding may be based on quality of food, but *Calanus pacificus* also apparently tends to pick large-sized particles when given a choice (Mullin, 1963; Richman and Rogers, 1969). In most selective feeding experiments with copepods it is not clear whether selection was active or due to increased feeding efficiency on larger particles as shown by Frost (1972). However, Richman and Rogers (1969) show that copepods can preferentially select large-sized particles. Selective feeding behavior of planktonic consumers has not been definitively studied at different concentrations of mixed prey, so functional responses cannot be described.

Perhaps the clearest documentation of preferential selective feeding is provided by the observations of Leong and O'Connell (1969) and O'Connell (1972) on the planktivorous northern anchovy (*Engraulis mordax*). The anchovy feeds either by filtering small particles from suspension or by seizing or biting individual large prey items. The relative importance of the two types of feeding was evaluated by presenting fish with mixtures of *Artemia* nauplii and adults in different proportions. Filter feeding behavior persisted only when *Artemia* adults comprised less than 7 percent of the available biomass of food. O'Connell showed that feeding by biting is the more efficient way for anchovy to obtain food when prey can be located visually and occur at a minimally dense concentration.

Numerical Response

Numerical response refers to the re'ationship between population growth of a consumer and density of its food; both individual and population growth will be considered under this topic. Like functional response, a hyperbolic type of curve describes the relationship between individual growth rate and density of food for a variety of marine and freshwater pelagic consumers (King, 1967; Paffenhöfer, 1970; Le-Brasseur, 1969; Reeve, 1970; O'Connell and Raymond, 1970). For a given species of consumer, the shape of the curve may vary depending on the type and size of food particles. At low food concentration adult *Calanus pacificus* feeds more effectively on relatively large species of net phytoplankton than on nannophytoplankton (Fig. 3). This may partly explain Paffenhöfer's (1970) observations on size of adult *Calanus* grown from eggs at different densities of monospecific cultures of algae. The species of algae may have had an effect, but larger *Calanus* also were obtained in cultures where larger-sized algal cells were used as food. In contrast, Mullin and Brooks (1970a,b) could not demonstrate that size of algal cells influenced either growth rate or size of the same species of *Calanus*. The amount and composition of lipid in *Calanus pacificus* are probably indirectly affected by the size of algal cell used as food (Lee et at., 1971) because the size of food particles determines ingestion rate.

Gross growth efficiency (K), the portion of ingested food which goes into growth of an individual during a time interval, appears to vary with density of food. Mullin and Brooks (1970a) found for *Calanus* that K decreased with increasing density of food. Kerr (1971a) reviews similar results for fish and points out that K must first increase before it levels off or decreases with increasing food density (Reeve, 1963). For cases where K decreases with increasing food density, Paloheimo and Dickie (1966) concluded that the rate of decrease depends on the size of food particles, being less for larger particles. Kerr (1971b) incorporated this effect into a model to demonstrate that large pelagic consumers will grow most efficiently (have high K values) on food particles which are large relative to the body size of the consumer; for large consumers the density of food particles had less influence on K than the size of food particles. So far, laboratory experiments have not revealed this effect of prey size on gross growth efficiency of a pelagic consumer (LeBrasseur, 1969; Mullin and Brooks, 1970a) but the failure may be due to limitations of laboratory experimentation, especially with regard to simulating natural foraging conditions (Kerr, 1971a). Variations in assimilation efficiency with prey density could partly explain the dependence of K on prey density, but no generalization can be drawn from the literature.

There is much less information available on population growth responses to prey density. King (1967) clearly demonstrated the effect of density of food particles on population growth of a planktonic rotifer, *Euchlanis dilatata,* in an initially unlimited environment. The instantaneous rate of population increase, r, plotted against density of food (unicellular algae) yielded a hyperbolic curve; a form of equation (2) may represent this hyperbolic relationship between food density and population growth rate (Caperon, 1967). The maximal r in King's experiments was obtained only at very high densities of food. Similarly, Hamilton and Preslan (1970) found that the maximal population growth rate of a marine ciliate was attained at unnaturally high densities of bacterial food. Comparable studies are not available for larger pelagic forms. Hall (1964) determined exponential population growth rates of a species of *Daphnia* at three food levels, but his two larger food densities were so high that the corresponding values of r were probably already on the asymptotic part of a hyperbolic curve.

King (1967) also found that the type (species) of algae fed to *Euchlanis dilatata* affected its population growth rate. The rotifer seemed to ingest three species of algae with equal facility so that differences in food quality or assimilability of the algae might account for observed variations in population growth. Generally, at a given density of food the instantaneous rate of population increase of the rotifer varied inversely with the size of the food particles. Arnold (1971) found significant differences in the population growth rate of *Daphnia pulex* depending on the species of algae used as food; size and shape of food particles may, however, be confounded with food quality in this investigation.

In sum, theoretical considerations suggest that consumers should grow best and increase in numbers most rapidly when utilizing food particles that are near the upper size limit of the range of particles which the consumers can handle. The fact that successful laboratory culture of larger particle grazers, e.g., *Calanus*, has come with the use of large-sized algae as food tends to support this. Nevertheless, much additional refined experimental work is required to demonstrate that particle grazers follow optimal feeding strategies.

Abundance and Distribution of Food in Nature

Conover (1968) uses the phrase "nutritionally dilute" to describe the environment of zooplankton because food particles are widely dispersed and many laboratory studies of feeding rates indicate that planktonic consumers achieve maximal rations only at what appear to be unnaturally high densities of prey. Discrepancies may arise when attempts are made to match estimated metabolic demands and food-collecting capacities of consumers to estimated food levels in the sea (Jorgensen, 1966). Obviously, error in one or more of these estimates could lead to a mismatch. For example, the filtering capacity of species of *Calanus* was grossly underrated in the past (see Paffenhöfer, 1971; Corner et al., 1972), and this was probably due partly to design of earlier feeding experiments and partly to the widespread use of small algal cells as food. *Calanus*, in contrast to the smaller *Pseudocalanus* for example, is probably adapted to feeding on the larger size of phytoplankton cells, the net phytoplankton. This is indicated by the results of Mullin and Brooks (1970a) and Paffenhöfer (1970), whose laboratory-cultured *Calanus*, grown at very low food densities of large-sized diatoms or dinoflagellates, were approximately the same size as field animals. Thus it seems risky to judge whether species of *Calanus* can meet their metabolic demands by comparison with levels of particulate organic matter in the sea (Butler et al., 1970). Despite all of the laboratory experiments and field observations that have been made on species of *Calanus*, we still do not have a clear conception of how these grazers utilize suspended particles in nature. Do they blunder through the ocean indiscriminately filtering and ingesting all living and nonliving particles which they happen to encounter, or do they follow optimal feeding strategies which cannot be brought out in simple experimental systems? Further, to what extent does patchiness of phytoplankton influence the feeding behavior and population growth of grazers? These questions cannot be answered, but perhaps some directions for future investigation can be indicated.

Estimates of average concentration of biomass of food in the ocean may be poor indicators of the amount of food available to consumers, even if it is assumed that the consumers can efficiently utilize the entire particle-size spectrum of food. The patchiness problem becomes increasingly critical to consumers as food density decreases; much of the world ocean may turn out to consist of widely dispersed prey patches separated by regions where food densities are near or below starvation level (Isaacs, 1965). Ivlev (1961) demonstrated that increasing overdispersion of food particles promotes greater ingestion rates of fish over their rates of feeding on a uniform distribution of the same number and type of food particles. Presumably this is because fish visually locate patches of prey. Often, groups of individuals of single prey species alternate in a linearly stratified fashion in the gut of planktivorous oceanic salmon (Allen and Aron, 1958; Ueno, 1968), suggesting that salmon encounter and feed upon swarms of prey, but also search-image predation could be involved.

For nonvisual particle grazers, estimates of average density of food will be meaningful if these consumers do not perceive the heterogeneous distribution of their food and do not actively search for and remain with patches containing high densities of prey. In this case the numerical response of consumers will be explicable on the basis of average densities of their food. This interpretation probably applies to observations that immature stages of herbivorous copepods are most abundant in subsurface patches of phytoplankton (Anderson et al., 1972; Mullin and Brooks, 1972).

Phytoplankton patchiness will constitute a "grainy" environment for a planktonic herbivore only if the herbivore encounters zones of significant size with phytoplankton densities greater than average. That is, the scale of patchiness must be considered relative to the size and motility of the herbivore. For a female of *Calanus pacificus*, for example, the relevant scale of patchiness would be on the order of a meter to perhaps a few 10's of meters. Because there is virtually no data on this scale of patchiness in phytoplankton and no theory on the size and permanence of patches, the influence of small-scale patchiness on feeding behavior of planktonic consumers is impossible to evaluate.

The effect of phytoplankton patchiness on individual growth rates of grazers may be evaluated indirectly. Populations of *C. pacificus* in

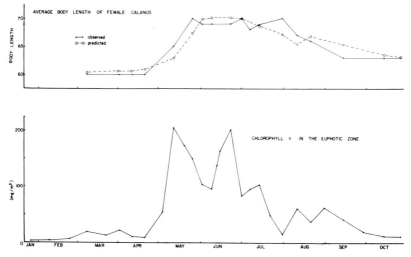

Figure 6. Seasonal variations in body (prosome) length of adult females of *Calanus pacificus* (above) and chlorophyll *a* in the euphotic zone (below) of the main basin of Puget Sound during 1965. Length units are ocular micrometer units (1 unit = 0.04 mm). Predicted lengths of *Calanus* were found by regression analysis of length measurements and estimated chlorophyll *a* in the euphotic zone during the generation time of *Calanus*. Generation times at different seasons were assumed to be determined solely by temperature and were estimated using the data of Mullin and Brooks (1970a,b). Unpublished chlorophyll data of G. C. Anderson and K. Banse.

the main basin of Puget Sound, Washington, do not seem to be unusually successful at utilizing low concentrations of phytoplankton. In 1965 extensive observations of phytoplankton standing stocks and primary production rates were made by G. C. Anderson and K. Banse. Zooplankton samples were also collected. Groups of adult females of *C. pacificus* were removed randomly from these samples and measured. Average body length of adult females is predictable from average abundance of chlorophyll in the euphotic zone during a generation (Fig. 6). When estimates of average body length of *Calanus* are plotted against estimates of phytoplankton carbon, a hyperbolic relationship emerges (Fig. 7); using Figure 3 as a basis for comparison, the largest adult *Calanus* occur in Puget Sound during periods of relatively high carbon concentration. Since the method for estimating plant carbon in Figure 7 errs in the direction of underestimation, it appears that phytoplankton patchiness contributes little to increased feeding efficiency of *Calanus* at low food concentrations or the curve would be displaced toward the left. This conclusion is tentative because the den-

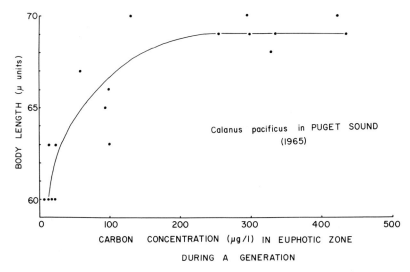

Figure 7. Relationship between body (prosome) length of adult females of *Calanus pacificus* and estimated phytoplankton carbon in the euphotic zone of the main basin of Puget Sound during 1965. Phytoplankton carbon was estimated using a seasonally variable carbon/chlorophyll ratio.

sity of phytoplankton carbon in laboratory cultures is not strictly comparable to estimates of phytoplankton carbon in the ocean. Usually most of this carbon in the ocean is not concentrated in the net plankton but in the nannoplankton.

These ideas can be evaluated only when methods are developed which will allow determination of precisely how much and what type of food particles grazers utilize under natural conditions. Fortunately, these methods are appearing. For example, Lee et al. (1971) describe an interesting relationship involving quantity and composition of lipid in *Calanus* which may specify the amount of food available and type of food eaten by copepods during development.

Acknowledgment: This paper is Contribution Number 748 from the Department of Oceanography, University of Washington. My research was supported by NSF grants GA-25385, GB-20182, and GA-31093 and by AEC contract AT (45-1)-2225, TA 26 (ref. RLO-2225-T26-8). I thank G. C. Anderson, K. Banse, J. Ambler, and P. Evans for providing unpublished data.

74   THE BIOLOGY OF THE OCEANIC PACIFIC

Literature Cited

Adams, J. A., and J. H. Steele. 1966. Shipboard experiments on the feeding of *Calanus finmarchicus* (Gunnerus), In: *Some Contemporary Studies in Marine Science*, pp. 19-35, H. Barnes, ed. George Allen and Unwin, London.

Allen, G. H., and W. Aron. 1958. Food of salmonid fishes of the western North Pacific. U. S. Fish. Wildl. Serv., Spec. Scient. Rept., Fish. No 237: 1-11.

Anderson, G. C., B. W. Frost, and W. K. Peterson. 1972. On the vertical distribution of zooplankton in relation to chlorophyll concentration In: *Biological Oceanography of the Northern North Pacific Ocean*, pp. 341-345, A. Y. Takenouti, ed. Idemitsu Shoten, Tokyo.

Arnold, D. E. 1971. Ingestion, assimilation, survival, and reproduction by *Daphnia pulex* fed seven species of blue-green algae. Limnol. Oceanogr., *16:* 906-920.

Beklemishev, C. W. 1962. Superfluous feeding of marine herbivorous zooplankton. Rapp. P. v. Cons. Int. Explor. Mer, *153:* 108-113.

Brooks, J. L., and S. I. Dodson. 1965. Predation, body size and composition of the plankton. Science, *150:* 28-35.

Burns, C. W. 1968a. Direct observations of mechanisms regulating feeding behavior of *Daphnia* in lakewater. Int. Revue ges. Hydrobiol., *53:* 83-100.

Burns, C. W. 1968b. The relationship between body size of filter-feeding Cladocera and the maximum size of particle ingested. Limnol. Oceanogr., *13:* 675-678.

Burns, C. W. 1969. Particle size and sedimentation in the feeding behavior of two species of *Daphnia*. Limnol. Oceanogr., *14:* 392-402.

Butler, E. I., E. D. S. Corner, and S. M. Marshall. 1970. On the nutrition and metabolism of zooplankton. VII. Seasonal survey of nitrogen and phosphorus excretion by *Calanus* in the Clyde Sea area. J. Mar. Biol. Ass. U. K., *50:* 525-560.

Caperon, J. 1967. Population growth in micro-organisms limited by food supply. Ecology, *48:* 715-722.

Conover, R. J. 1966. Factors affecting the assimilation of organic matter by zooplankton and the question of superfluous feeding. Limnol. Oceanogr., *11:* 346-354.

Conover, R. J. 1968. Zooplankton—life in a nutritionally dilute environment. Am. Zoologist, *8:* 107-118.

Corner, E. D. S., R. N. Head, and C. C. Kilvington. 1972. On the nutrition and metabolism of zooplankton. VIII. The grazing of *Biddulphia* cells by *Calanus helgolandicus*. J. Mar. Biol. Ass. U. K., *52:* 847-861.

Davids, C. 1964. The influence of suspensions of microorganisms of different concentrations on the pumping and retention of food by the mussel (*Mytilus edulus* L.). Netherlands J. Sea Res., *2:* 233-249.

Engelmann, M. D. 1966. Energetics, terrestrial field studies and animal productivity. Adv. Ecol. Res., *3:* 73-115.

Frank, P. W. 1957. Coaction in laboratory populations of two species of *Daphnia*. Ecology, *38:* 510-519.

Frost, B. W. 1972. Effects of size and concentration of food particles on the feeding behavior of the marine planktonic copepod *Calanus pacificus* Limnol. Oceanogr., *17:* 805-815.

Gold, K. 1971. Growth characteristics of the mass-reared tintinnid *Tintinnopsis beroidea*. Mar. Biology, *8:* 105-108.

Greve, W. 1972. Ökologische Untersuchungen an *Pleurobrachia pileus* 2. Laboratoriums Untersuchungen. Helgoländer wiss. Meeresunters., *23:* 141-164.

Gulland, J. A. 1970. Food chain studies and some problems in world fisheries. In: *Marine Food Chains,* pp. 269-315, J. H. Steele, ed. Univ. Calif. Press, Berkeley.

Hall, D. J. 1964. An experimental approach to the dynamics of a natural population of *Daphnia galeata mendotae.* Ecology, *45:* 94-112.

Hall, D. J., W. E. Cooper, and E. E. Werner. 1970. An experimental approach to the production dynamics and structure of freshwater animal communities. Limnol. Oceanogr., *15:* 839-928.

Hamilton, R. D., and J. E. Preslan. 1970. Observations on the continuous culture of a planktonic phagotrophic protozoan. J. Exp. Mar. Biol. Ecol., *5:* 94-104.

Hardy, A. C. 1924. The herring in relation to its animate environment, part 1. Ministry of Agriculture and Fisheries, Fish. Invest., Ser. 2, 7(3).

Holling, C. S. 1959. Some characteristics of simple types of predation and parasitism. Canad. Entomol., *91:* 385-398.

Holling, C. S. 1965. The functional response of predators to prey density and its role in mimicry and population regulation. Mem. Entomol. Soc. Canada, No. 45.

Isaacs, J. D. 1965. Larval sardine and anchovy relationships. Calif. Coop. Oceanic Fish. Invest., *10:* 102-140.

Ivlev, V. S. 1961. *Experimental Ecology and Feeding of Fishes.* D. Scott, transl. Yale Univ. Press, New Haven.

Jorgensen, C. B. 1966. *Biology of Suspension Feeding.* Pergamon Press, Oxford.

Kerr, S. R. 1971a. Analysis of laboratory experiments on growth efficiency of fishes. J. Fish. Res. Bd. Canada, *28:* 801-808.

Kerr, S. R. 1971b. A simulation model of lake trout growth. J. Fish. Res. Bd. Canada, *28:* 815-819.

Kerr, S. R., and N. V. Martin. 1970. Trophic-dynamics of lake trout production systems. In: *Marine Food Chains,* pp. 365-376, J. H. Steele, ed. Univ. Calif. Press, Berkeley.

King, C. E. 1967. Food, age, and the dynamics of a laboratory population of rotifers. Ecology, *48:* 111-128.

LeBrasseur, R. J. 1969. Growth of juvenile chum salmon (*Oncorhynchus keta*) under different feeding regimes. J. Fish. Res. Bd. Canada, 26(6): 1631-1645.

LeBrasseur, R. J., W. E. Barraclough, O. D. Kennedy, and T. R. Parsons. 1969. Production studies in the Strait of Georgia. Part III. J. Exp. Mar. Biol. Ecol., *3:* 51-61.

Lee, R. F., J. C. Nevenzel, and G. A. Paffenhöffer. 1971. Importance of wax esters and other lipids in the marine food chain: phytoplankton and copepods. Mar. Biol., *9:* 99-108.

Leong, R. J. H., and C. P. O'Connell. 1969. A laboratory study of particulate and filter feeding of the northern anchovy (*Engraulis mordax*). J. Fish. Res. Bd. Canada, *26:* 557-582.

Loosanoff, V. L., and J. B. Engle. 1947. Effect of different concentrations of micro-organisms on the feeding of oysters (*O. virginica*). Fish. Bull., *51:* 29-57.

McAllister, C. D. 1970. Zooplankton rations, phytoplankton mortality and the estimation of marine production. In: *Marine Food Chains,* pp. 419-457, J. H. Steele, ed. Univ. Calif. Press, Berkeley.

McMahon, J. W. 1965. Some physical factors influencing the feeding behavior of *Daphnia magna* Straus. Canad. J. Zool., *43*: 603-611.

McMahon, J. W., and F. H. Rigler. 1963. Mechanisms regulating the feeding rate of *Daphnia magna* Straus. Canad. J. Zool., *41*: 321-332.

Monakov, A. V. 1972. Review of studies on feeding of aquatic invertebrates conducted at the Institute of Biology of Inland Waters, Academy of Science, USSR. J. Fish. Res. Bd. Canada, *29*: 363-383.

Mullin, M. M. 1963. Some factors affecting the feeding of marine copepods of the genus *Calanus*. Limnol. Oceanogr., *8*: 239-250.

Mullin, M. M., and E. R. Brooks. 1970a. The effect of concentration of food on body weight, cumulative ingestion, and rate of growth of the marine copepod *Calanus helgolandicus*. Limnol. Oceanogr., *15*: 748-755.

Mullin, M. M., and E. R. Brooks. 1970b. Growth and metabolism of two planktonic, marine copepods as influenced by temperature and type of food. In: *Marine Food Chains,* pp. 74-95, J. H. Steele, ed. Univ. of Calif. Press, Berkeley.

Mullin, M. M., and E. R. Brooks. 1972. The vertical distribution of juvenile *Calanus* (Copepoda) and phytoplankton within the upper 50 m of water off La Jolla, California. In: *Biological Oceanography of the Northern North Pacific Ocean,* pp. 346-354, A. Y. Takenouti, ed. Idemitsu Shoten, Tokyo.

O'Connell, C. P. 1972. The interrelation of biting and filtering in the feeding activity of the northern anchovy (*Engraulis mordax*). J. Fish. Res. Bd. Canada, *29*: 285-293.

O'Connell, C. P., and L. P. Raymond. 1970. The effect of food density on survival and growth of early post yolk-sac larvae of the northern anchovy (*Engraulis mordax* Girard) in the laboratory. J. Exp. Mar. Biol. Ecol., *5*: 187-197.

Paffenhöfer, G. A. 1970. Cultivation of *Calanus helgolandicus* under controlled conditions. Helgoländer wiss. Meeresunters., *20*: 346-359.

Paffenhöfer, G. A. 1971. Grazing and ingestion rates of nauplii, copepodids and adults of the marine planktonic copepod *Calanus helgolandicus*. Mar. Biol., *11*: 286-298.

Paloheimo, J. E., and L. M. Dickie. 1966. Food and growth of fishes. III. Relations among food, body size, and growth efficiency. J. Fish. Res. Bd. Canada, *23*: 1209-1248.

Parsons, T. R., and R. J. LeBrasseur. 1970. The availability of food to different trophic levels in the marine food chain. In: *Marine Food Chains,* pp. 325-343, J. H. Steele, ed. Univ. Calif. Press, Berkeley.

Parsons, T. R., R. J. LeBrasseur, and J. D. Fulton. 1967. Some observations on the dependence of zooplankton grazing on the cell size and concentration of phytoplankton blooms. J. Oceanogr. Soc. Japan, *23*: 10-17.

Parsons, T. R., R. J. LeBrasseur, J. D. Fulton, and O. D. Kennedy. 1969. Production studies in the Strait of Georgia. Part II. Secondary production under the Fraser River plume, February to May 1967. J. Exp. Mar. Biol. Ecol., *3*: 39-50.

Reeve, M. R. 1963. Growth efficiency in *Artemia* under laboratory conditions. Biol. Bull., *125*: 133-145.

Reeve, M. R. 1964. Feeding of zooplankton, with special reference to some experiments with *Sagitta*. Nature, *201*: 211-213.

Reeve, M. R. 1970. The biology of chaetognatha. I. Quantitative aspects of growth and egg production in *Sagitta hispida*. In: *Marine Food Chains,* pp. 168-189, J. H. Steele, ed. Univ. Calif. Press, Berkeley.

Reeve, M. R., and M. A. Walter. 1972. Conditions of culture, food-size selection, and the effects of temperature and salinity on growth rate and generation time in *Sagitta hispida* Conant. J. Exp. Mar. Biol. Ecol., *9:* 191-200.

Richman, S., and J. N. Rogers. 1969. The feeding of *Calanus helgolandicus* on synchronously growing populations of the marine diatom *Ditylum brightwellii.* Limnol. Oceanogr., *14:* 701-709.

Ryther, J. H. 1969. Photosynthesis and fish production in the sea. Science, *166:* 72-76.

Schindler, D. W. 1968. Feeding, assimilation and respiration rates of *Daphnia magna* under various environmental conditions and their relation to production estimates. J. Anim. Ecol., *37:* 369-385.

Schindler, J. E. 1971. Food quality and zooplankton nutrition. J. Anim. Ecol., *40:* 589-595.

Schoener, T. W. 1969. Models of optimal size for solitary predators. Amer. Natur., *103:* 277-313.

Sheldon, R. W., A. Prakash, and W. H. Sutcliffe, Jr. 1972. The size distribution of particles in the ocean. Limnol. Oceanogr., *17:* 327-340.

Slobodkin, L. B. 1954. Population dynamics of *Daphnia obtusa* Kurz. Ecol. Monogr., *24:* 69-88.

Sprules, W. G. 1972. Effects of size-selective predation and food competition on high altitude zooplankton communities. Ecology, *53:* 375-386.

Steele, J. H. 1972. Factors controlling marine ecosystems. In: *The Changing Chemistry of the Oceans,* pp. 209-221, D. Dyrssen and D. Jagner, eds. Nobel Symp. 20.

Sushchenya, L. M. 1970. Food rations, metabolism and growth of crustaceans. In: *Marine Food Chains,* pp. 127-141, J. H. Steele, ed. Univ. Calif. Press, Berkeley.

Ueno, M. 1968. Food and feeding behavior of Pacific salmon. I. The stratification of food organisms in the stomach. Bull. Japanese Soc. Sci. Fish., *34:* 315-318.

The Structure of Deep Benthic Communities from Central Oceanic Waters

Robert R. Hessler
Scripps Institution of Oceanography
La Jolla, California

> As we descend deeper and deeper in this region its inhabitants become more and more modified, and fewer and fewer, indicating our approach towards an abyss where life is either extinguished, or exhibits but a few sparks to mark its lingering presence. Its confines are yet undetermined, and it is in the exploration of this vast deep-sea region that the finest field for submarine discovery yet remains. (Forbes, 1859)

In THIS QUOTATION, Forbes summed up what today is the commonplace observation that as one turns his attention from shallow-water benthos toward the greater depths, he witnesses a gradual but continuous alteration in the fauna. The species composition changes, as does to a lesser extent the taxonomic composition at the level of genus, family, and so on. The standing crop decreases markedly. The diversity of the community changes. We suspect that many aspects of the organism's physiology and life cycle change as well.

Where do these changes end? Most detailed studies concentrate on the coastal deep sea, rendering it unlikely that they have encompassed the entire gradient. A few studies, particularly that of the Soviet's Institute of Oceanology, indicate that the end of the gradient is reached only upon penetration to the central cores of the major oceans. Here the water is at its deepest, except for the trenches, which should be regarded as special cases. Here the chances of continental influence are at a minimum, and being central gyre waters, surface conditions are at their least productive. It is here that one witnesses Forbes' "lingering presence" in its purest form. The benthic fauna of this region is the subject of this paper.

I will deal only with soft-bottom communities. Although animals have been scraped off deep-sea hard bottoms, no one has attempted a serious analysis of hard-bottom communities. Indeed, most deep-sea benthic biologists avoid hard bottoms, which to them are usually nothing more than graveyards for their gear.

The primary concern of this paper will be the structure of the community, since almost nothing is known of rates in the ocean depths. Our ignorance in this area stems from two causes. First, because of the constancy of the environment, there is little detectable periodicity in life cycles. Even at relatively modest bathyal depths, most organisms show no seasonality (Sanders and Hessler, 1969; Scheltema and Sanders, 1971; Rokop, in preparation), although Schoener's (1968) study of ophiuroids may demonstrate an exception to this. Thus, there are none of the usual time-related clues that allow one to determine recruitment rate, growth rate, or life span, and, of course, tagging is still impossible.

The second limitation on rate studies is that, although great advances are being made, it is still extremely difficult to obtain healthy animals for physiological study. Most of what is brought to the surface is either dead or moribund. Even where there are no obvious signs of damage, the physiological health of the animals may be questioned, since invariably they have experienced major changes in temperature or pressure during their capture (Pamatmat, in press). The *in situ* study of metabolism is still in its infancy, although it promises real success (Smith, in press).

There are few sources of information on deep benthic communities far from continental influence. The occasional isolated, semisuccessful samples taken by major expeditions as they crossed the open ocean are of little help. The greatest effort in community studies has been that of the Institute of Oceanology of the USSR (Zenkevich, 1969). Of particular use in the present context are their maps of benthic biomass and studies of trophic composition. The Gay Head-Bermuda Transect, studied from the Woods Hole Oceanographic Institution, has included stations in the Sargasso Sea (Sanders, Hessler, and Hampson, 1965; Hessler and Sanders, 1967; Sanders and Hessler, 1969). Finally, I have initiated studies of the benthic community under central gyre waters in the Pacific Ocean, particularly the North Pacific (Hessler and Jumars, in preparation).

Faunal Composition

At the grossest taxonomic level, the faunal composition of deep-sea, soft-bottom communities is much the same as that in shallow water.

However, at lower taxonomic levels, the differences increase, until at the level of species there are few or no similarities. At the same time, the proportional importance of various taxa changes. These changes are most pronounced at the break between continental shelf and slope, which therefore defines the edge of the deep sea. But even within the deep benthos, shifts can be documented. Table 1 compares the relative abundances of deep-sea taxa in central gyre waters of the North Atlantic and North Pacific to inshore deep sea in the North Atlantic. The 4,000-meter dividing line is used because that is water depth below the Gulf Stream, which is a good separator of marginal and central oceanic waters. On the deep side of this boundary, polychaetes, although still the dominant element, decrease in importance, while the peracarid crustaceans become proportionally more common. Within the peracarids, amphipods become rarer, while isopods and especially tanaids become increasingly significant (Sanders, Hessler, and Hampson, 1965). Bivalve Mollusca become less important.

The ratios of abundance of various higher taxa are surprisingly conservative at abyssal depths, being essentially the same in central gyre waters of the North Atlantic and North Pacific. Polychaetes, tanaids, bivalves, and isopods (in descending order of importance) make up about 85 percent of the macrofauna.

These conclusions are based on the animals caught in grabs or small-scale trawls, and screened with a .297 mm or .420 mm mesh. Furthermore, only those taxa traditionally regarded to be macrofauna are considered. Rowe and Menzel (1971) have quantitatively documented the fact that on an average, individuals are much smaller in the deep sea than in shallow water, a point which has often impressed workers on a wide variety of taxa. Thus, in reality most of the animals taken in central gyre waters are actually meiofaunal in size, that is, they would pass through a 1.0 mm screen.

Individuals of certain taxa are routinely so small that they are all of meiofaunal size. Members of this category commonly seen in deep-sea samples are Foraminifera, Nematoda, Ostracoda, and harpacticoid Copepoda. To assess abundance of these groups properly, a 40-60μ mesh screen should be used. A single sample from the North Pacific gyre, washed with a 62μ screen yielded an order of magnitude greater abundance of "meiofaunal taxa," with no significant increase in the number of "macrofaunal taxa" over the yield of a 297μ screen. Yet, even with the coarse 297μ screen used in the North Pacific study, these meiofaunal taxa appeared in large numbers. Considering nematodes, copepods, and ostracods combined, there were 2.26 times as many individuals in these categories as in the combined macrofaunal taxa. As in shallower waters, there was a great preponderance of nematodes, fol-

Table 1.   The percent composition of macrofaunal taxa of deep-sea, soft-bottom
communities

Taxonomic group	Northwest Atlantic[1]		Northcentral Pacific[2]
	< 4,000 m	> 4,000 m	5,600 m
	%	%	%
PORIFERA	<.1	.2	1.1
CNIDARIA	.5	.5	1.4
POLYCHAETA	70.4	55.6	54.4
OLIGOCHAETA	.7	......	2.1
SIPUNCULIDA	5.8	4.6	.4
ECHIURIDA	<<.1	......	.4
PRIAPULOIDEA NEMERTINA POGONOPHORA	.9	......	......
TANAIDACEA	1.6	19.3	18.1
ISOPODA	1.0	12.2	5.9
AMPHIPODA	4.1	1.5	......
CUMACEA MISC. ARTHROPODA	.1	.2	......
APLACOPHORA	.6	.3	1.1
BIVALVIA	13.0	4.3	7.0
GASTROPODA	.3	.6	.4
SCAPHOPODA	.5	.2	2.4
OPHIUROIDEA	.3	.8	.7
ECHINOIDEA	.1	.2	......
CRINOIDEA ASTEROIDEA HOLOTHUROIDEA	.3	......	.4
ECTOPROCTA	>.4	......	2.1
BRACHIOPODA	......	......	.7
ASCIDIACEA	<<.1	......	1.1

[1] Data for the Northwest Atlantic come from anchor dredge samples on the Gay Head-Bermuda Transect (Sanders, Hessler, and Hampson, 1965). The column for <4,000 meters averages ten stations ranging in depth from 200 meters to 2,870 meters. Seven stations having a depth range of 4,436-5,001 meters were used for the second column.

[2] The Northcentral Pacific data is an average of ten 0.25 m² cores, all from the same spot at 28°30'N, 155°20'W at 5,497-5,825 meters depth.

lowed by copepods, and then ostracods (Tietjen, 1971; Thiel, 1971). In contrast to their numerical abundance, the biomass of meiofaunal taxa would comprise only a fraction of that of macrofaunal taxa at bathyal depths (Wigley and McIntyre, 1964; Thiel, 1972). The nature of this relationship is still unknown for bottoms beneath midocean gyres.

Coastal deep-water transects show that Foraminifera become increasingly important with depth (Tietjen, 1971). Although foraminiferal tests were abundant and diverse at the North Pacific gyre study area, it is still impossible to assess their relative abundance because of our inability to judge whether most of the tests were alive upon capture. At those depths, most foraminiferans are large and arenaceous. Thus, even when stained, it is difficult to discern the presence of protoplasm. In spite of this difficulty, the importance of Foraminifera in the community cannot be doubted. They display a diversity of 20 to 50 species in 0.25 m² samples. In the case of the few species whose living state can be discerned, typically several living individuals will occur in a single sample of 0.25 m². Such abundance within single species is unknown among macrofaunal taxa.

Perhaps related to the foraminiferans are small organisms composed of dendritically joined or anastomosing organic tubules with clay particles agglutinated to the surface (Fig. 1). Mention of these is completely absent from the literature, yet they comprise the bulk of organic remains in the samples from the central North Pacific. For the most part, these organisms have been ignored in the sorting of deep-sea samples in the past, although they are characteristic components. The Soviets call them "vitvistii kamochki," or "little branching clusters" (A. Kuznetsov, personal communication). They may be protozoans called Xenophyophoria (Tendall, 1972), although they are orders of magnitude smaller than any thus far reported. Unfortunately, as with the Foraminifera, we have been unable so far to distinguish living from dead remains, and therefore cannot derive an accurate assessment of their relative importance.

The benthic microfauna of Central gyre waters remains completely unknown.

The megafauna of Central gyre waters is poorly understood because of the great technical difficulties of sampling this size category at such great depths. Sedentary animals are so sparse that they are rarely collected, and mobile forms, which are equally sparse, easily avoid the collecting device. As a result, major taxa may still have escaped our attention, as the recent discovery in this region of a truly gigantic amphipod has shown (Fig. 2C) (Hessler, Isaacs, and Mills, 1972).

The sparsity of megafauna is documented by the rarity of their appearance in the many deep-water camera surveys. Yet baited camera studies demonstrate that the large, mobile scavengers are capable of concentrating their effects on one area, given the proper stimulus, such

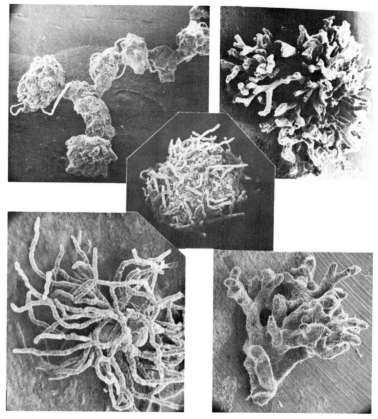

Figure 1. "Kamochki" from the North Pacific gyre, 5,500 meters depth. Scanning electron photomicrographs (magnification 50-75 times).

as food (Fig. 2A,B) (Dayton and Hessler, 1972). If large organic parcels are a significant portion of the energy input to such areas, these large, mobile scavengers would be instrumental in disseminating this energy over the bottom.

Standing Crop

The faunal lists of the *Challenger* Expedition, when tabulated according to depth, gave rise to the truism that in the deep sea, standing crop is inversely correlated with depth of water and distance from major land masses (Murray, 1895). Interestingly, the *Challenger* data

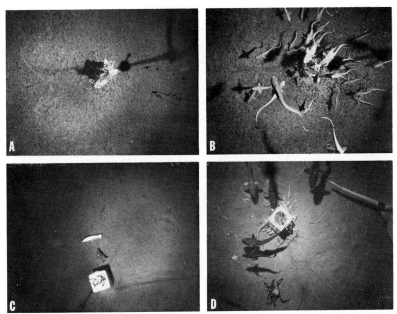

Figure 2. Studies of animals attracted to bait. A,B: Northwestern Pacific, 34°03'N, 163°59'E, 5,850 meters; in A, the bait has just reached the bottom; B is 25 hours later. C: central North Pacific gyre, 23°54.5'N, 144°04.9'W, 5,300 meters; the white amphipod is about 280 mm long. D: Northeastern Pacific, off Baja California, 30°53'N, 116°45'W, 2,000 meters; after 12 hours, the bait has attracted an asteroid, lithodid crab, skate, macrurid and lycodid fish, and numerous *Hyalinoecia stricta* (quill worms).

could hardly be considered proof of this relationship, because quantitative samples were never taken.

The first legitimately quantitative information was accumulated in the 1950's by the Soviets (Vinogradova, 1962) and the Danes (Spärck, 1951). The continuing Soviet studies (Filatova, in Zenkevich, 1969), coupled with the work of a variety of other programs (Hartman and Barnard, 1958; Wigley and McIntyre, 1964; Carey, 1965; Sanders, Hessler, and Hampson, 1965; Griggs, Carey, and Kulm, 1969; Sanders, 1969; Rowe and Menzel, 1971; Tietjen, 1971; Thiel, 1971; Rowe, 1971) reveal a relatively simple pattern which can be applied to all oceans.

The data collected on the Gay Head-Bermuda Transect for macrofaunal taxa (Sanders, Hessler, and Hampson, 1965) show this pattern most clearly. On the outer continental shelf and upper continental slope the standing crop is at a maximum, yielding densities of

5,000-22,000 individuals per square meter. Going down onto the lower continental slope (>500 m) and upper continental rise, faunal density drops precipitously to fewer than 1,000 ind./m². On the lower continental rise and abyssal plain, both beyond the Gulf Stream and in the Sargasso Sea, standing crop drops to as low as 33 ind./m². Up the Bermuda Slope there is a slight increase in standing crop, but the densities are much lower than at comparable depths on the continental slope. Thus, standing crop diminishes by more than two orders of magnitude with increasing depth and distance from land.

The accepted cause for such patterns is the amount of available food (Sanders and Hessler, 1969; Filatova, in Zenkevich, 1969; Thiel, in press). Food comes from the euphotic zone or from terrestrial runoff. As the food settles to the bottom, it diminishes in amount because of autolysis, bacterial decay, or scavenging by midwater organisms. Thus, the amount of food reaching the bottom is inversely related to depth. The importance of this factor is shown on the continental slope, where stations at different depths are so close together that the surface waters above them must be similarly productive, yet where benthic standing crop and water depth are closely coupled.

Because primary productivity is highest in continental coastal waters, where depth is constant the benthic standing crop will decrease with increasing distance from land. This is shown by the Sargasso Sea samples, where depth varies only a few hundred meters, yet standing crop is correlated with distance from land. The importance of the geographical proximity to primary productivity is illustrated even better in comparisons of open ocean samples from under the productive Equatorial upwelling to samples from the unproductive Central gyres (Filatova, in Zenkevich, 1969) (Fig. 3A). Although the depth is the same, the benthic standing crop under the former is an order of magnitude higher than under the latter.

Thus the map of standing crop in the deep benthos matches the map of surface productivity (Fig. 3B), except that coastal gradients are amplified by correlated changes in water depth. As a result, the areas of lowest standing crop are under the centers of the major oceanic gyres. These are the areas farthest from land, with lowest surface productivity, and usually with the greatest depths. (Trenches are excluded in the present considerations because their great depth combined with their frequent proximity to major land masses yields confused relationships.)

In the gyre areas total density of macrofaunal taxa ranges from 30 to 200 ind./m² (Sanders, Hessler, and Hampson, 1965; Hessler and Jumars, in preparation), with biomasses of 0.05-0.01 gms (wet wt.)/m² (Filatova, in Zenkevich, 1969). The agreement in figures for

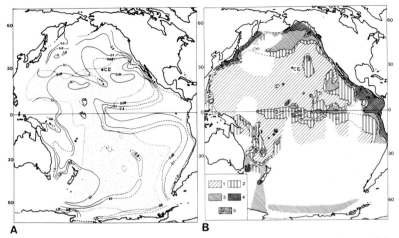

A   B

Figure 3.   The Pacific Ocean. A: Benthic standing crop in grams wet weight
per square meter (after Filatova, in Zenkevich, 1969); the stipled area is the
oligotrophic zone as determined by Sokolova (in Zenkevich, 1969). B: Primary
productivity of the water column (after Koblents-Mishke, 1965); values for
1-5 are 100, 100-150, 150-250, 250-650, and >650 mg carbon/m²/day, respectively.
The asterisk marked by C II is the location of the box core samples discussed in
the text.

numerical density between the Sargasso Sea and the Central North
Pacific gyre is quite good, reflecting their general similarity in surface
productivity and water depth.

Diversity

Another idea arising from the *Challenger* results was the belief
in a lower diversity of species in deep-sea communities. The intuitive
feeling was that surely the extreme environmental conditions (low
temperature, high pressure, no light, little food) would be intolerable
to all but a modest collection of life forms.

This idea was soundly disproved by samples taken along the
Gay Head-Bermuda Transect (Hessler and Sanders, 1967). Large,
unwinnowed samples from depths up to 4,700 m yielded diversities
(including length of the species list and equitability) far higher than
those of the shallow, inshore communities of the transect region, and
equivalent to diversities of shallow-water, soft-bottom communities of
the tropics.

Sanders (1968) incorporated these findings into a hypothesis
which emphasizes the stability and predictability of the environment as

a controlling factor for faunal diversity. According to this view, if environmentally stable conditions persist for a long enough period of time, speciation and immigration will cause species diversity to increase gradually, as species in the community gradually become biologically accommodated to each other. Dayton and Hessler (1972) have disputed the mechanism whereby this process occurs, but the importance of environmental predictability remains unchallenged.

There yet lingered, however, the possibility that in those areas where standing crop is very low by virtue of the small amount of food entering the system, the diversity of the community is decreased, there being a lower limit to the density of individuals that will allow members of a species to interact for reproductive purposes. If this were the case, it is in the deep, benthic communities beneath the central gyres that one would expect to see it. The data from the North Pacific disallow this possibility. Summing the fauna from ten box cores collected in 1969 within five miles of each other (Fig. 3), we obtain a large enough collection of animals to allow comparison of diversity to that of other regions. Figure 4 plots the diversity of combined polychaetes and bivalves to the rarefaction curves for that faunal segment plotted by

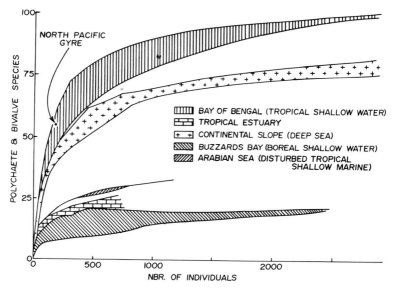

Figure 4. Envelopes of rarefaction curves for polychaetes and bivalves (after Sanders, 1968), and showing the diversity of summed box core samples from one spot under the North Pacific gyre.

Sanders (1969). The low number of individuals per species places the combined samples on that portion of the curve where diversity is highly dependent on sample size. Still, the diversity in the Central North Pacific gyre (176 ind./54 species) is clearly extremely high, ranking along with samples from the shallow-water tropics and higher than samples from the the North Atlantic continental slope.

Since these data come from a community about as food-poor as this planet has to offer, it suggests that sparseness of food in itself does not affect diversity. It is interesting to speculate on how small, sparse, relatively immobile deep-sea species avoid the reproductive problems resulting from low numerical density. It is possible that under these stable environmental conditions the life span is sufficiently long that even a low rate of intraspecific encounters is adequate. Increased chance for encounter may be another selective advantage to partitioning the standing crop of a species into parcels of small size. Refined chemical sensory abilities must also play an important part.

Trophic Structure

Most investigators of deep benthic communities have concluded that deposit feeding is the overwhelmingly dominant trophic type (Menzies, 1962; Sokolova, 1965; Sanders and Hessler, 1969). However, there is one potentially important exception to this. Sokolova (1965), working with the extensive series of samples collected by the Institute of Oceanology from the Pacific and Indian Oceans, distinguishes between the communities of "eutrophic" and "oligotrophic" bottoms. As the name implies, eutrophic benthic communities receive greater quantities of food and thus have higher standing crops. In oligotrophic areas, where food supply is low, the standing crop is minimal. Oligotrophic bottoms are ones of low sedimentation rate, tending to be deeply oxidized red clays, often with manganese nodules. Shark teeth and squid beaks are not uncommon. There is less than 0.25 percent organic carbon in the sediment. Obviously, deep central gyre bottoms fall well within the definition of the oligotrophic zone (Fig. 3A).

According to Sokolova, in oligotrophic waters the community is dominated by suspension feeders (>70%) rather than deposit feeders (<6%). Carnivores are also less important in oligotrophic regions than in eutrophic regions. Sokolova suggests that organic material accumulates so slowly in oligotrophic regions that the deposited organic carbon is too refractory to be nutritious. In her view, suspended organics, while less abundant than in eutrophic waters, are nevertheless proportionately more nutritious in the oligotrophic, when compared to deposited organics.

Sanders and Hessler (1969) did not find a major difference in trophic structure between the oligotrophic Sargasso Sea and the relatively richer Northwest Atlantic coastal waters. The same general taxa occurred in both areas, and in all cases deposit feeders dominated. Similarly, the fauna of box core samples from the Central North Pacific gyre is heavily dominated by deposit feeders. More recently, Sokolova (in Zenkevich, 1969; 1972) has modified her position to allow that deposit feeders dominate throughout the meiofauna. Her definition of meiofauna includes essentially all that we obtain using our collecting procedures. Thus, the discrepancy in our findings is reconciled.

The conclusion that there are trophic-type differences between eutrophic and oligotrophic bottoms is based primarily on the megafauna, that is, large animals which are so sparse that they can be obtained only with large trawls. Among the suspension feeders this includes sponges, arcid and pectinid bivalves, serpulid polychaetes, and brachiopods. Among the large deposit feeders would be several families of asteroid, holothurian, and echinoid echinoderms, and protobranch bivalves.

Baited camera studies demonstrate how the eutrophic-oligotrophic relationship applies to the most highly selective of the deposit feeders, the scavengers. In coastal deep water, a variety of animals are attracted to bait. Fish, prawns, octopuses, crabs, polychaetes, ophiuroids, and asteroids are the most common visitors (Fig. 2D). In the more sterile deep water far from land, bait still yields a quick response, but only by fish, natantian decapods, and amphipods, all of which are strong swimmers (Fig. 2B, C). Ordinarily widespread on the bottom, these forms are quick to sense and concentrate on the large parcels of food which sporadically appear. Such parcels become more infrequent going into the open ocean benthos.

Under such circumstances only the most vagile types can profit from such sources. Ambulatory megafaunal species would usually arrive at the food too late. If larger organic parcels form a significant portion of the diet of megafaunal scavengers, then one would expect selective reduction of ambulatory forms from oligotrophic gyre communities. The mobile scavengers which remain would generally not be caught by deep trawling techniques, and therefore their importance would not be appreciated.

The reduction of other megafaunal deposit feeders will have different explanations, but all will revolve around reduction of available food. Perhaps polychaetes and protobranch bivalves are rarer in oligotrophic trawl samples because the general diminution of individuals in these groups makes them unavailable for sampling by that technique. Nonselective deposit feeders, such as holothurians, may drop out because this inefficient mode of feeding does not yield sufficient energy where the percent of labile organic carbon is so low.

Conclusions

The purpose of the present paper has been a description of benthic communities of the deep, central oceanic gyres. By means of comparisons to the coastal deep sea and shallow inshore waters, several changes were noted. Standing crop decreases seaward. Among the major macrofaunal taxa are shifts in relative proportions, with a clear decrease in the importance of polychaetes, amphipods, and bivalves, accompanied by an accentuation of tanaids and isopods. Among the megafauna there is a profound reduction in deposit-feeding forms, with only natantian scavengers persisting to a major degree. Except for shifts among the macrofaunal taxa, these changes are thought to be caused by the markedly decreasing food supply, and it is likely that even the shift in macrafaunal taxa will prove to stem from this cause.

It must be remembered, however, that these changes are within a larger framework of basic uniformity. The deep-sea fauna is a coherent entity, the largest biotope on the face of the planet. It is recognizable at relatively high taxonomic levels, and while species vary geographically, taxa are basically cosmopolitan at levels as low as the genus. The community can also be characterized by dominance of the deposit-feeding habit and by high species diversity in terms of both species number and equitability. While it is still too early to make definitive statements, we are confident that deep-sea benthonts will display special patterns of reproduction, recruitment, life span, and so on. This uniformity of a deep-sea fauna stems from the basic similarity of the physical regime throughout its extent. Of greatest importance are temperature, light, conditions of sedimentation, currents, pressure, and most important of all, the great temporal stability which characterizes all aspects of the deep-sea environment.

Acknowledgments: This work was supported by grants GB 14488 and GA 31344X of the National Science Foundation. The baited camera studies result from the generous cooperation of the Marine Life Research Group, under the direction of Prof. John D. Isaacs. I am grateful to Peter A. Jumars for sorting the polychaetes from the Central North Pacific, and for his criticism of the manuscript. This paper is a contribution of Scripps Institution of Oceanography.

Literature Cited

Carey, A. G. 1965. Preliminary studies on animal-sediment interrelationships off the central Oregon coast. Ocean Sci. & Ocean Eng., *1:* 100-110.

Dayton, P. K., and R. R. Hessler. 1972. Role of biological disturbance in maintaining diversity in the deep sea. Deep-Sea Res., *19:* 199-208.

Forbes, E. 1859. *The Natural History of the European Seas.* John Van Voorst, London.

Griggs, G. B., A. G. Carey, and L. D. Kulm. 1969. Deep-sea sedimentation and sediment-fauna interaction in Cascadia Channel and on Cascadia Abyssal Plain. Deep-Sea Res., *16:* 157-170.

Hartman, O., and J. L. Barnard. 1958. The benthic fauna of the deep basins off southern California. Allan Hancock Pacific Exped., *22:* 1-67.

Hessler, R. R., J. D. Isaacs, and E. L. Mills. 1972. Giant amphipod from the abyssal Pacific Ocean. Science, *175:* 636-637.

Hessler, R. R., and H. L. Sanders. 1967. Faunal diversity in the deep sea. Deep-Sea Res., *14:* 65-78.

Koblents-Mishke, O. I. 1965. Primary production in the Pacific. Okeanology (Engl. transl.), *5:* 104-116.

Menzies, R. J. 1962. On the food and feeding habits of abyssal organisms as exemplified by the Isopoda. Int. Revue ges. Hydrobiol., *47:* 339-358.

Murray, J., 1895. A summary of the scientific results. Challenger Repts., 1608 pp.

Pamatmat, M. M. Benthic community metabolism on the continental terrace and in the deep sea in the North Pacific. Int. Revue ges. Hydrobiol., in press.

Rowe, G. T. 1971. Benthic biomass and surface productivity. In *Fertility of the Sea,* Vol. 2, pp. 441-454, J. D. Costlow, ed. Gordon & Breach, New York.

Rowe, G. T., and D. W. Menzel. 1971. Quantitative benthic samples from the deep Gulf of Mexico with some comments on the measurement of deep-sea biomass. Bull. Mar. Sci., *21:* 556-566.

Sanders, H. L. 1968. Marine benthic diversity: a comparative study. Amer. Nat., *102:* 243-282.

Sanders, H. L. 1969. Benthic marine diversity and the stability-time hypothesis. Brookhaven Symposia in Biology, Diversity and Stability in Ecological Systems, *22:* 71-81.

Sanders, H. L., and R. R. Hessler. 1969. Ecology of the deep-sea benthos. Science, *163:* 1419-1424.

Sanders, H. L., R. R. Hessler, and G. R. Hampson. 1965. An introduction to the study of deep-sea benthic faunal assemblages along the Gay Head-Bermuda Transect. Deep-Sea Res., *12:* 845-867.

Scheltema, R. S., and H. L. Sanders. 1971. Reproduction and population dynamics of some protobranch bivalves from the continental shelf, slope, and abyss of the northeastern United States. 6th Eur. Symp. Mar. Biol., Rovinj, Yugosl., Sept. 27-Oct. 2, 1971. Abstr. to be published in Thalassia.

Schoener, A. 1968. Evidence for reproductive periodicity in the deep sea. Ecology, *49:* 81-87.

Smith, K. L. Deep-sea benthic community respiration: an *in situ* study at 1850 meters. Science, in press.

Sokolova, M. N. 1965. The uneven distribution of food groupings of the deep water benthos in relation to uneven sedimentation. Okeanology (Engl. transl.), *5:* 85-92.

Sokolova, M. N. 1972. Trophic structure of deep-sea macrobenthos, Mar. Biol., *16:* 1-12.

Spärck, R. 1951. Density of bottom animals on the ocean floor. Nature, *168:* 112-113.

Tendall, O. S. 1972. A monograph of the Xenophyophoria. Galathea Rept., *12:* 7-103.

Thiel, H. 1971. Häufigkeit und Verteilung der Meiofauna im Bereich des Island-Färöer-Rückens. Berichte Deut. Wissensch. Komm. f. Meeresforsch., *22:* 99-128.

Thiel, H. 1972. Die Bedeutung der Meiofauna in küstenfernen benthischen Lebensgemeinschaften verschiedener geographischer Regionen. Verhandlungsber. Deut. Zool. Gesellsch., 65th Jahresversamm.:37-42.

Thiel, H. The trophic structure of the deep-sea benthos. Int. Revue ges. Hydrobiol, in press.

Tietjen, J. H. 1971. Ecology and distribution of deep-sea meiobenthos off North Carolina. Deep-Sea Res., *18:* 941-957.

Vinogradova, N. G. 1962. Some problems of study of deep-sea bottom fauna. Jour. Oceanogr. Soc. Japan, 20th Anniv.:724-741.

Wigley, R. L., and A. D. McIntyre. 1964. Some quantitative comparisons of offshore meiobenthos and macrobenthos south of Martha's Vineyard. Limn. & Oceanogr., *9:* 485-493.

Zenkevich, L. A., ed. 1969. Deep-Sea Bottom Fauna, Pleuston. In: *The Pacific Ocean, The Biology of the Pacific Ocean,* Vol. 7, part 2 pp. 1-353, V. G. Kort, chief ed.

Fishery Potential from the Oceanic Regions

Brian J. Rothschild
National Oceanic and Atmospheric Administration
National Marine Fisheries Service
Southwest Fisheries Center
La Jolla, California

THIS PAPER HAS a scope as vast as the oceanic province itself. Not only is the oceanic province extremely large, but relative to the various other regions of the ocean, very little is known about the resources that occupy its waters. Because so little is known about the oceanic province, inferences on this region tend to be based upon the more complete knowledge of the shallower coastal areas. In addition, the myriad of species and the vastness of the area require that a brief exposition such as this be treated with a rather broad brush. Thus, while my original intent was to treat the North Pacific Ocean, it appeared from the level of knowledge on oceanic resources and the degree of generality and exposition which conform to the style of these Colloquia that the discussion might better be served by a review of the fishery resources of the oceanic province in general. Rather than run through the traditional list of oceanic species or attempt to devise fixed estimators of the potential oceanic fishery production, I shall concentrate upon a few concepts which are used to determine this production, namely extrapolation from trends in present catches and food chain dynamics.

Estimation of the magnitude of resources of the oceanic province involves questions of practical as well as academic importance. Fish are an important element in world commerce, and they are likely to become more important in the future. Fish are also a particularly important commodity to the developing countries, perhaps not so much as a nutritional additive, but as a source of raw material which can be utilized to stimulate economic growth, thus contributing to international stability. This increase in the significance of fishery resources will be accompanied by (1) a reduction in the rate of catch increase for conventionally harvested species and in some instances decreases in the catch; (2) increases in national jurisdiction over fishing stocks that will reduce free access to them; (3) a rise in costs associated with conventional fishing technology; and (4) an increased demand for fishery products.

From the practical point of view we must ask how we can make the best use of our fishery resources, whether they be coastal resources or oceanic resources. There are of course many alternative paths to the development of unused resources and to the "wiser use" of those fishery resources which are already utilized. These alternative paths toward development of fishery resources exist in both developing and developed nations and in coastal waters as well as oceanic regions. The best alternative path for fishery development will depend to a large extent upon the physical capability of the various resources to produce sustained yields. This point is evidenced by noting that the present annual world catch of marine fish is about 60 million metric tons per year. If we anticipate the maximum sustainable catch to be about 80 million metric tons, our developmental strategy would be considerably different than if we anticipate the maximum sustainable catch to be 200 million metric tons. It is, therefore, important that the analysis of the maximum sustainable catch be made with great care and responsibility because appropriate development and utilization of our fishery resources are essential parts of our general resource problems.

In order to guess at the future production from the oceanic region, it is useful to attempt to extrapolate from present trends in world catches in general. These oceanic resources include the tunas, the billfishes, the squids, the flyingfish, the dolphin, and a myriad of various deepwater and surface species. The trends in catches have been reviewed by Gulland (1971). He observed a rather steady increase, at a rate of 7 percent per year, in the world catch. The world catch in 1938 was 21×10^6 tons, in 1956 it was about 30 million tons, and in 1970 it was nearly 70 million tons. The increase of 7 percent per annum has, of course, been compounded by declines and increases in individual fisheries as well as the institution of new fisheries and the elimination of others.

Gulland (1971) points out that in terms of new fisheries we have the following:

Peruvian anchovetta	(1955)	60,000 tons	(1961)	5 million tons
Norwegian mackerel	(1963)	20,000 tons	(1967)	870,000 tons
Thailand otter trawl	(1962)	78,000 tons	(1965)	337,000 tons
Southeast Atlantic hake	(1962)	100,000 tons	(1966)	410,000 tons

Off our own Pacific coast we have had some dramatic increases in catch of hake and other groundfish, and there is considerable potential increase in the catch of some fish such as the anchovy. On the other hand, we have had declines in some fisheries in the North Pacific such as the famous sardine decline, king crab in the Subarctic Pacific, and sauries off Japan. The yellowfin sole peaked at 500,000 metric tons in 1960 (estimated population 500,000 metric tons in 1964). Some fisheries have declined elsewhere, such as the herring in the Atlantic. Because of these rather short-term fluctuations it is rather dangerous to extrapolate from catch trends in individual stocks. Nevertheless we can make some interesting generalizations about the increases in catch:

1. The world catch continues to increase.

2. New fisheries must be found to maintain the increases in total catch.

3. Since relatively few individual fisheries have become extinct (possibly owing to a damping off in fishing effort as the stocks decline in density), we must be fishing an increasing number of species.

4. Fishing changes the productivity of stocks by changing mortality rates, and one can speculate that greater total yields might be obtained from many overfished stocks than from a few moderately fished stocks.

If the increase of 7 percent per year were to continue, we would hit somewhat more than 100 million tons (of conventional forms) by 1980 and about 800 million tons by the year 2000. Most authors feel that the final limit is of a magnitude closer to 100 million tons than to 800 million tons. If we can proceed, however, to several hundred million tons per year, we will almost certainly be exploiting nonconventional fisheries such as squid, larger zooplankton, and small oceanic fish. We would almost certainly need to consider harvesting significant quantities of the oceanic forms if it were possible to do so.

The Pacific Ocean has traditionally supplied somewhat less than half the total world catch, and there is no reason to doubt that it will continue to do this. For the total world catch to exceed 100 million tons and the Pacific to go beyond about 50 million tons, there will need to be considerable reliance on the nontraditional types of fish. Before these fish can be caught, it is almost certain that new technologies will have to be developed. In fact, these technologies are probably a major constraint upon the harvest of oceanic fishes. The potential catches of the various oceanic forms have been summarized by Gulland (1971) from extrapolation and other evidence (Table 1.).

It is interesting to note that the total scombrid catches, primarily the large fish, have remained stable over the last several years at about 1.5×10^6 to 1.6×10^6 tons, so large increases in these forms (with the

Table 1. Summary of potential catches from oceanic resources

Type of fish	Tons
Whales	
Large baleen (1,900 Blue Whale Units)	1,640,000
Sperm whales (25,000 animals)	500,000
Small whales	500,000
Dolphins, porpoises	?
Salmon	
North Pacific	500,000
Atlantic	15,000
Tunas	
Large tunas	
Pacific	350,000 – 450,000
Atlantic	200,000 – 250,000
Indian Ocean	100,000 – 150,000
Skipjack	
Pacific	500,000 – 800,000
Atlantic	250,000 – 300,000
Indian Ocean	160,000 – 300,000
Other small tunas	
Frigate mackerel	(1,000,000)
Bonito	(500,000)
Little tuna	?
Thynnus tonggol	?
Sharks	(500,000)
Coryphaenids	(1,000,000)
Squids	10-100 million
Myctophids, etc.	Hundreds of millions
Red crab	(1,000,000)

Note: Salmon are included here because most of their growth is accomplished in open oceans. (From: *The Fish Resources of the Ocean.* FAO, 1971, J. A. Gulland, ed. Fishing News (Books) Ltd.).

exception of the skipjack tuna) are unlikely. Also, of the important tunas of commerce, roughly 65 percent are caught in the Pacific Ocean, 10 percent in the Indian Ocean, and 25 percent in the Atlantic Ocean. Very roughly, 50 percent of the world ocean is the Pacific, 30 percent the Atlantic, and 20 percent the Indian, which suggests only on the basis of surface area that larger catches may be expected from the Indian Ocean.

If we are to have really large increases in world catch, they are likely to come from the oceanic regions. We have so little experience with the oceanic regions that there is little basis for extrapolation, and this raises the question of using an alternative procedure to guess the potential yield of fish from the oceanic region. The best known alternative procedure is to estimate the potential production of an oceanic region on the basis of primary production and the production at each successive stage in the food chain.

The classic food chain argument is extremely simple. I shall reiterate it here so that I can base my further remarks on certain of its aspects. The food chain can be viewed as a collection of factories. This is diagrammed in Figure 1. An extremely important aspect of this scheme is the number and configuration of the factories. The real "road map" could be quite complicated with such things as multiple tracks, switching yards, and tracks that pass at least once through the factory of origin before reaching the factory of destination. Furthermore, each factory has been treated essentially as a "black box," measuring only the inputs and the outputs without considering the internal working mechanism. (There are, of course, a number of papers that consider internal working mechanisms of the factories, but most of the literature on food chain estimation of production tends to be unconcerned with the details of the phenomena that occur in the "factories." Indeed, the system may, in many respects, be less sensitive to the workings of the factories than to the number of factories.)

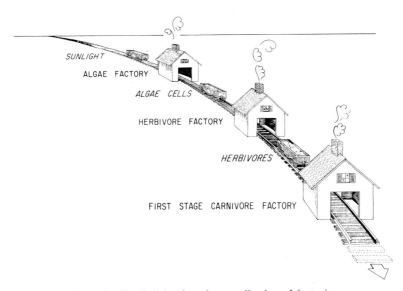

Figure 1.   Food chain viewed as a collection of factories.

This food chain approach has actually been used to deduce the total yield of the ocean and portions of the ocean such as the oceanic region. Ryther (1969) discusses the primary productivity of the various ocean regions, the rate of production of the "algae factory." First there is the open ocean or the *oceanic province* which is our major concern today. This province occupies 90 percent of the world ocean. However, its mean primary productivity is rather low (about 50 grams of carbon per square meter per year is fixed in organic matter). Next we have the *coastal zone,* which represents about 10 percent of the total ocean area including offshore areas of high productivity. The productivity is higher here, about 100 grams of carbon per square meter per year. Finally we have the *upwelling areas,* which occupy a fraction of a percent of the total area and have the highest productivity of all—300 grams of carbon per square meter per year. It is significant to note that, even with its low primary production per square meter, the oceanic province represents, according to Ryther's statistics, about 85 percent of the total annual productivity of the sea, yet only 2 or 3 percent of the world fish catch is taken from this area.

Next we need to evaluate the transformation process in each factory. How much material is carried in the boxcars to the next factory and how much is blown out of the smokestacks as metabolites and other material? In discussing the transformation process we will discuss the food chain approach as defined by Ricker (1969). Ricker discusses the two relevant coefficients:

E, the ecotrophic coefficient—the fraction of a prey species' annual production that is consumed by predators (trophic referring to nutritive or food levels);
K, the growth coefficient—the predators' annual increment of weight divided by the quantity of food they have consumed.

Ricker (1969, p. 94) adjusts the ecotrophic coefficient for recycling and arrives at the following values:

	Growth coefficient (K)	Ecotrophic coefficient (with recycling adjustment) (E)
Primary consumption (grazing on green plants)....	15%	66%
Higher levels	20%	75%

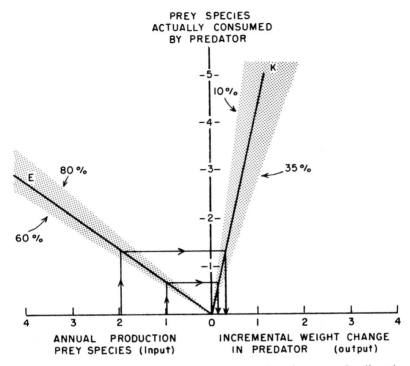

Figure 2. The relation between the annual production of prey species (input), the prey species actually consumed by the predation, and the incremental change in the production of the predator (output). The two lines which trace the transformation of input to output demonstrate the effect of an incremental change of input on output.

giving *KE* values of roughly 10 percent at the herbivore stage and 15 percent at the higher levels. Figure 2 is a diagram of this process, showing that changes in the production of prey species can have proportionately equal effects on the change in incremental weight of the predators.

Note that by the simple way the problem is formulated, the relation between the ordinate values and the abscissa values must be straight lines passing through the origin, a situation which may be quite unlikely in the real world. This is, of course, true of all constant transfer coefficients that are given in the literature.

Ryther used coefficients, *KE,* of 10, 15, and 20 percent for the oceanic, coastal, and upwelling provinces respectively. The exact magnitude of these coefficients is not at all certain. It is also not clear that they differ in a consistent way between the oceanic, coastal, and up-welling provinces. They may, in fact, be more variable depending on the level of the chain rather than the location of the chain. There is, however, general agreement that the coefficients lie between 10 and 20 percent.

Next we need to know how many factories or levels exist in our simple model. Ryther assumed that harvestable fish production would occur at the fifth trophic level in the oceanic province, at the third in the coastal zone, and between the first and second (1 1/2) in the upwelling zone. This assignment along with the efficiencies provided Ryther with a guess at the total potential production of fish in the world ocean of about 240 million metric tons. His figures give equivalent total fish production to the coastal and upwelling provinces, but allocate 1/150 of the total potential fish production to the oceanic province. In other words, only 10 percent of the world ocean accounts for something like 99 percent of the total potential fish production. These particular re-sults of Ryther's have been criticized in the literature (see Alverson et al., 1970). The criticism boils down to the fact that many plausible alternate coefficients could have been used to obtain strikingly different conclusions.

Why is there a considerable difference in fish production deduced from impressions of the populations (Table 1) and the food chain analysis? For example, the food chain method gives a fish production in the oceanic province of only 1.6 million metric tons. The actual catch (not the production) is actually approaching this quantity, and yet there are rather large stocks as judged by the number of larvae of un-exploited scombrids in the oceanic regions, not to mention the squids, deepwater fishes, and others.

Since it is difficult to pinpoint the difficulties, we might examine the model itself. The most sensitive spot in our food chain model is the length of the food chain. To take an example we note that

$$P_n = P_o k^n$$

where P_n is the production at the nth stage, P_o is the primary produc-tion, k is the transfer coefficient, and n is the number of stages. Table 2 gives values of P_n for several P_o at various k and n. Clearly, the num-ber of trophic levels, n, affects P_n very strongly. Changes in k appear to be relatively unimportant for most reasonable values of k. Using the above equation for particular areas may give us some idea about the adequacy of the classic food chain model in the regions of interest. As we explore the details of the simple model, we are aware that the di-vergence between the food chain approach and the extrapolated esti-mate of oceanic productivity could very well be produced by the model

Table 2. Expected levels of fish production in the oceanic province (derived from food chain model)

Level of primary production	Transfer coefficients (k)								
	20%			15%			10%		
	Trophic levels (n)								
	2	3	5	2	3	5	2	3	5
gCm⁻²yr⁻¹									
50	10	2	0.08	7.5	1.13	0.025	5	0.5	0.005
100	20	4	0.16	15	2.25	0.051	10	1	0.01
300	60	12	1.48	45	6.75	0.153	30	3	0.03

diverging from the real world, a difference in the number of trophic steps which was used for the food chain model, or extrapolations that are too high. It is less likely that the values of the transfer coefficients or the estimates of primary productivity produce these divergences. With respect to realism it is well known that it is extremely difficult to evaluate at which trophic level an animal actually resides. While it is relatively easy to determine what an animal eats, it is not easy to determine what its food ate; in any case, no organism is going to fit into neat integral trophic levels, especially during the course of its life.

Given that the factory model is a very simplified abstraction, we must ask whether it is a satisfactory approximation. It is, in the sense that we can come up with almost any answers by modifying the various coefficients. But this may not be a good criterion. The main utility of such a model will almost certainly lie in enabling us to understand how the fundamental processes in the ocean *differ* from the simple model. Most critical are the pathways and feedback mechanisms that channel the flow of energy among the factories and the inner workings of the factories themselves. Short circuiting of the chain can produce substantial differences in production of animals of harvestable size.

Some of the problems of generalizing about the biology of oceanic fishes with respect to trophic dynamics could be considered in terms of "average" coefficients and trophic levels. It may be that certain species which inhabit particular water masses (see Ebeling, 1967) are characterized by particular sets of coefficients and trophic levels and that the differences in these features among water masses would induce more variability in these characteristics than that which would obtain, for example, for demersal fishes. In addition the "two-layered" tropical ocean produces some intriguing problems in food chain dynamics with respect to introducing nutrient-rich water into the photic zone by eddies downstream of islands (see Barkley, 1972) and by the diurnal migration of some fish between the surface and deep layers.

Another intriguing problem in oceanic food chains, which again involves the question of how we interpret trophic levels, is the role of organic substances and the equilibrium between dissolved and particulate organic material. Exactly how this material is incorporated into the food chain and if it is incorporated in significant amounts remains elusive. Provasoli (1963) has compiled a considerable amount of information on this subject. He emphasizes the difficulties of dealing with organic substances in sea-water which are caused by their minute concentrations: "Characterization of organic compounds in sea-water is complex because the dissolved organic C averages 2 mg/s (maxima up to 20 mg/s). These minimal quantities have to be separated from 35,000 mg of inorganic salts in a liter of sea-water. . . ." These tiny quantities of organic material evidently contribute to the existence of the phenomenon that Provasoli calls "good" and "bad" waters; for example, the ". . . productivity along the coast of California is far less than around the British Isles, yet the phosphate content of California waters is many times higher."

Even though modifications in the transfer function may not be relatively important, it is worth emphasizing the simplicity of the model by pointing out that it does not consider time lags. This is easily demonstrated by examining simple control system equations. The transfer function that is typically used in food chain dynamics is the zero order function:

$$I = k\Omega$$

where I is the output, Ω is the output, and k is a constant. There are quite plausible, more complex functions which can represent the process, viz., the first-order and second-order differential equations:

$$I = k\Omega + L\frac{d\Omega}{dt}$$

and

$$I = k\Omega + L\frac{d\Omega}{dt} + M\frac{d^2\Omega}{dt^2}.$$

A simple unit-forcing function can, depending upon the value of the constants, L and M, generate quite different results for the time-behavior of the output of the system. This is shown in Figure 3.

Thus, with the zero order equation the quantity of material produced at a factory is a constant fraction of the material that enters the factory. With more complex, more realistic equations, there are time

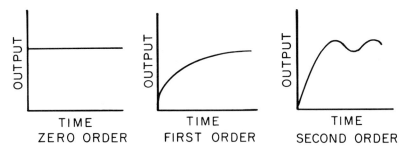

Figure 3. Typical outputs resulting from constant input as a function of time showing different kinds of lag effects possible for zero-order, first-order, and second-order differential equations.

lags which could cause considerable oscillations in the system. (Are these oscillations reflected in varying year-classes of fish?) Another time-related effect is the age effect. Most population theory relates to $dN/dt = zN$, and it can be shown that the average age of the organisms in the population is $1/z$. However, the average age of the population can be modified by changes in the predator population or by fishing. If we assume that size is a function of age, and that diet is a function of size, then the transfer coefficient must be continually changing. The point of this is that the zero order equation is conventionally used, but the higher order equations are much more likely to operate. If, for example, the damping ratio is small, the fluctuations can have yearly or quarterly periods, and the trophic position of organisms must be changing constantly with mortality and other vital rates.

Other factors that affect the nature of the transfer coefficient are various changes in the environment that can be temporary but occur at important stages during the organism's life, and changes of a longer term. Take, for example, the whole question of nutrition (see Phillips, 1969). It is conceivable that there are long-term and short-term changes in the amounts of proteins, fats, and carbohydrates present in a trophic level. The essential amino acids and fats, the sparing action of certain amino acids and fats, and mineral and vitamin requirements may also vary. As another example, consider the fluctuations in temperature that can operate upon an animal's temperature tolerance, preference, "appetite," and digestion. When do these variations of nutrition, temperature, and a host of other factors create significant perturbations in the system? We do not know.

Thus, it is clear that there are many unanswered questions concerning the use of the factory model to forecast the productivity of the oceanic region. Perhaps we need to ask what additional knowledge the food chain model will produce. Perhaps we need to look at some newer configurations of the problem involving the fate and residence time of packets of energy in groups of animals. We could in this context consider a variety of queuing questions for each group of animals: what is the arrival time of packets of energy at different densities and nutritional quality, what are the lengths of the queues, what are the holding times, and what is the effect of queue impatience?

All of this has simply served to point out that the oceanic province is about as little known, with respect to the kinds of information required to make resource decisions, as it was at the time of the *Challenger* expedition. If economic pressures accelerate the harvesting of the oceanic regions, then we need to concentrate upon obtaining information that will promote correct decisions with respect to the exploitation of these resources.

Literature Cited

Alverson, D. L., A. R. Longhurst, and J. A. Gulland. 1970. How much food from the sea. Science, *168:* 503-505.

Barkley, Richard A. 1972. Johnston Atoll's Wake. J. of Mar. Res., *30*(2): 201-216.

Ebeling, Alfred W. 1967. Zoogeography of tropical deep-sea animals. Studies in Tropical Oceanography (Miami), *5:* 593-613.

Gulland, J. A., ed. 1971. *The Fish Resources of the Ocean.* Published by arrangement with FAO by Fishing News (Books) Ltd., Surrey, England.

Phillips, Arthur M., Jr. 1969. Nutrition, digestion and energy utilization. In: *Fish Physiology,* Vol. 1, Excretion, ionic regulation and metabolism, pp. 391-432, W. S. Hoar and D. J. Randall, eds. Academic Press, New York.

Provasoli, L. 1963. Organic regulation of phytoplankton fertility. In: *The Sea,* Vol. 2, pp. 165-219, M. N. Hill, ed. Interscience Publ., New York.

Ricker, William E. 1969. Food from the Sea. In: *Resources and Man: A Study and Recommendations,* pp. 87-108. W. H. Freeman and Co., San Francisco.

Ryther, J. H. 1969. Photosynthesis and fish production in the sea. Science, *166*(3901): 72-76.

Enzymatic Adaptations to Deep Sea Life

P. W. Hochachka
Department of Zoology
University of British Columbia
Vancouver, B. C., Canada

Since the advent of evolutionary theory, environmental adaptations have been charted at many levels of biological organizations—at physiological, anatomical, and behavioural levels as well as at the most basic biochemical level. Least understood of these are what we term "biochemical adaptations to the environment." In part, this lack of information is an historic outcome of the development of adjacent fields of science and of an appropriate technology. But a more important reason for the hiatus in this field stems from the nature of the subject itself: biochemical adaptations usually are not apparent macroscopically, hence the biologist may be aware neither of their existence nor of their mechanisms.

To illustrate what I mean, consider the apparent macroscopic attributes of two hypothetical species of fishes—one, a surface species, living at the top of the water column at high temperatures and low pressures; the second, an abyssal species, living at low temperatures and high pressures. In much of their anatomy, physiology, and even behaviour the two species may be much alike. Both display comparable hydrodynamic properties and hence comparable swimming velocities; both display comparable growth and comparable osmoregulatory abilities. Above the biochemical level we may find little reason to regard one species as, say, barophyllic and stenothermic, and the other as barophobic and eurythermic. Only when we begin to "dissect" the biochemical machinery of the two species do we learn that the two organisms can conduct the same basic functions at comparable rates despite widely disparate physical conditions because of the existence of basic biochemical differences between them. It is indeed the biochemical adaptations to their respective environments which makes the two species so similar in the things they can do.

Further, as I shall illustrate in this paper, biochemical adaptations can largely be grouped into two categories: those which are "compensatory" and those which are "exploitative." Compensatory adaptations are essentially homeostatic in nature and can be looked upon as corrective or restorative mechanisms. Thus, the organism responds to any environmental parameter which adversely affects biochemical reaction rates by harnessing mechanisms for returning those rates to their previous levels.

Other biochemical changes may give the organism a wholly new potential for making use of its environment or for invading a new environment. We term such changes "exploitative adaptations." In contrast to biochemical adjustments which are compensatory or restorative, exploitative adaptations are, strictly speaking, unnecessary. The species could survive without these new potentials, but with them it may do significantly better in its native habitat and, more importantly, it may enter a new habitat previously unavailable to it.

In this context, I shall examine potential enzyme mechanisms available to organisms during adaptation to position in the sea. In particular, I will consider three environmental factors that are often associated with "position" in the marine environment; namely, temperature, pressure, and oxygen availability. I shall begin by considering the manners in which organisms have adapted to one of the most critical environmental parameters: the oxygen available to the cell.

Adaptations to Oxygen

Physiological limitations on oxygen availability

Throughout most of the water column, the environmental availability of oxygen is always adequate to support the routine metabolism of gill-breathing animals. This is true for both vertebrates and invertebrates. However, it is also true that the oxygen requirements of metazoans can be satisfied only through the use of accessory oxygen delivery systems. The design of these systems varies widely from the low pressure circulation systems of many invertebrates to the high pressure circulation systems of the vertebrates. Under a variety of stressful conditions ("burst" activities of various forms), the rate of oxygen delivery cannot meet the total oxygen demands of all the tissues. This is particularly true, of course, for air-breathing, diving vertebrates. Because of their mode of respiration and because of the depth of diving (during which lungs and trachea are collapsed), these vertebrates enter an environment which for practical purposes is devoid of available oxygen. The oxygen delivery systems in all vertebrates, but particularly in diving forms, then, are "weak-link" systems, and important biochemical problems arise from this limitation.

The vertebrate solution to this limiting situation is the regulation of blood circulation to favour some tissues (the heart and the central nervous system) at the expense of peripheral tissues in general, and skeletal white muscle in particular. An impressive capacity has evolved in vertebrate muscle as a consequence for extracting energy by anaerobic mechanisms. We shall consider these mechanisms in general before considering how they are specifically adapted by marine airbreathing vertebrates to allow diving excursions of great depth and great duration.

Glycogen as an energy source for anaerobic muscle function

During limiting oxygen conditions, glycogen serves as the primary carbon and energy source in all vertebrates. Glycogen mobilization is initiated by the enzyme glycogen phosphorylase, which is strategically positioned between the fuel depot and the enzyme machinery required for its anaerobic degradation. Because of its pivotal position, it is not surprising that glycogen phosphorylase is under tight hormonal, ionic, and metabolite control (see, for example, Drummond, 1971).

Activation of glycogen phosphorylase can be triggered hormonally (by epinephrine, norepinephrine, glucagon, etc.) and ionically by Ca^{++}. Both activators lead to the formation of an active tetrameric enzyme, phosphorylase a, from an inactive dimer, phosphorylase b, but their modes of action differ somewhat. Hormonal activation is indirect and complex. The primary action of epinephrine is the activation of adenyl cyclase, which catalyzes the reaction,

$$ATP \rightarrow 3', 5' \; AMP \; (cyclic \; AMP) + PP_i$$

Cyclic AMP in turn activates protein kinase by dissociating an inactive complex containing catalytic and regulatory subunits to an active protein kinase, free of its regulatory subunits. Protein kinase catalyzes the first of two phosphorylation reactions which lead to the dimerization of phosphorylase b, forming the active a tetramer. Deactivation of these enzymes is accomplished by specific phosphatases.

Regulation by Ca^{++} involves a direct activation of phosphorylase kinase (Figure 1). With the arrival of an electric impulse and the depolarization of the sarcolemma, Ca^{++}, which in the resting state is sequestered in the sarcoplasmic reticulum, is liberated. The free Ca^{++} ac-

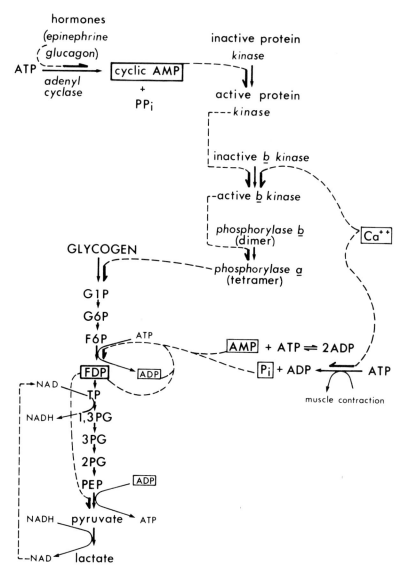

Figure 1.   The control of glycogen mobilization in vertebrate muscle. Activation of reaction steps is indicated by thick arrows.

tivates b kinase as shown in Figure 1. It also activates myofibrillar ATPase and thus is seen to play a dual and pivotal role in coupling muscle metabolism and muscle contraction. These events lead to a controlled and efficient release of glucose residues from glycogen and set the stage for a glycolytic activation which in its extent and in its precision (over a tenfold change in rate occurring within a minute) far exceeds that of all other tissues of the vertebrate body.

Enzymatic basis for the glycolytic potential of muscle

Enzyme concentrations. When compared to tissues such as the liver, brain, or kidneys, vertebrate skeletal muscle displays the highest levels of glycogen phosphorylase, hexokinase (HK), phosphofructo-kinase (PFK), phosphoglycerate kinase (PGK), and lactate dehy-drogenase (LDH), as well as other glycolytic enzymes. Recent studies by Kemp (1971), for example, comparing liver and muscle PFK, in-dicate specific activities about one hundred fold greater in muscle than in liver. Similar though smaller differences are documented for other glycolytic enzymes (Scrutton and Utter, 1968). Such high enzyme ac-tivities correlate well with the high glycolytic capacities of skeletal muscle. High enzyme levels do not of themselves account for two im-portant characteristics of muscle glycolysis which are not usually ob-served in other tissues: (1) its high capacity for function in the gly-colytic direction; and (2) its capacity for sudden changes in glycolytic rate. The basis for these characteristics is to be found not in the amount of any given enzyme but rather *in the kind of enzyme* (the isozyme type) catalyzing each of the glycolytic reactions.

Tissue-specific isozymes. Many and probably all of the glycolytic enzymes in muscle occur in isozymic forms which are fairly specific to that tissue (see, for example, Annals N.Y. Acad. Sci., Vol. 151, Arti-cle 1). Detailed documentation of their "adaptedness" for function in the microenvironment of muscle cells cannot be made for all of these forms (nor have I the space to examine all of them in this article); however, entirely convincing arguments can be made for some of them. The muscle isozyme of PFK is one of these. It, along with PGK and pyruvate kinase (PK), catalyzes reactions which proceed with large free energy drops: the PFK step proceeds with a $\triangle G° = -3.4$ kcal/mole; PGK proceeds with a $\triangle G° = -4.5$ kcal/mole; PK pro-ceeds with a $\triangle G° = -7.5$ kcal/mole. Thermodynamically, these reac-tions are essentially irreversible and they therefore "poise" the flux of carbon through the multienzyme pathway in the catabolic direction. In a sense, these three reactions are preadapted for "valvelike" control function. Indeed, in most organisms and tissues thus far examined, these three enzymes are identified as the major control sites in glycolysis. Of the three, PFK usually plays the predominant role in glycolytic activation, probably because it is the first "committed" enzyme step in the glycolytic path.

Glycolytic activation during muscle contraction and the central role of PFK

As indicated in Figure 1, the ATP cleavage which occurs during muscle contraction initiates a cascade pattern of glycolytic activation. Central to this event is the overall energy charge of the cell. Concomitant with the fall in ATP levels, rising ADP, AMP, and P_i levels potently activate muscle PFK by increasing PFK affinity for F6P and by reversing ATP inhibition of the enzyme. Moreover, several additionally critical components to PFK activation occur at this time (Figure 2):

1. F6P reverses any residual ATP inhibition of the enzyme by reducing ATP affinity of the allosteric site on the enzyme. Thus F6P increases the $K_i(ATP)$ but does not affect the $K_m(ATP)$.

2. F6P substrate saturation follows sigmoidal kinetics; thus this compound is both a substrate and a positive modulator, as the binding of the first F6P molecule facilitates binding of subsequent ones. (Parenthetically, it should be stressed that ATP is also a substrate and a modulator of PFK; however, it is a negative modulator, binding at a specific allosteric site on the enzyme).

3. Functionally, the most important aspect of the glycolytic activation occurring during muscle contraction is *product activation of PFK*. As shown in Figures 1 and 2, both FDP and ADP product-activate PFK, and more than any other single factor contribute to the exponential, "flare-up" nature of glycolytic activation in muscle. Similar glycolytic activation does not occur in other tissues; hence, it is not surprising that this regulatory property is *either lost or is less pronounced* in PFKs from tissues such as liver, red blood cells, or intestine.

4. The final aspect of PFK catalysis which contributes to the precision of PFK control in muscle concerns enzyme-ligand affinities. Thus, muscle PFK displays a two- to threefold greater affinity for its substrate (F6P) than do PFKs from other tissues; under physiological conditions, it will therefore compete for limiting F6P with a two- to threefold greater ability. Similarly, high enzyme-substrate affinities are known for other glycolytic isozymes specific to muscle (aldolases, pyruvate kinases, hexokinases); all of these presumably contribute to the overall efficiency of carbon flow in the glycolytic direction. In addition, muscle PFK shows higher affinities for key regulatory metabolites, such as the adenylates, than do PFK forms from other tissues, a

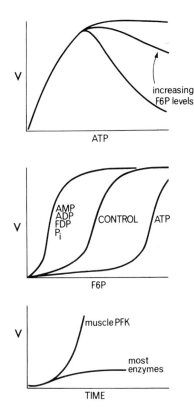

Figure 2. Control of muscle phospho-fructokinase. In the upper panel, ATP is seen to be both a substrate and an inhibitor of the reaction. The binding of ATP occurs at two sites: the cata-lytic site and the allosteric site. F6P reverses binding at the allosteric site, hence reverses inhibition, but it does not affect the binding of ATP at the catalytic site. In the middle panel, ATP is seen to greatly reduce the affinity of PFK for F6P. In contrast, the positive modulators (AMP, ADP, FDP, and P_i) reverse this effect of ATP and increase the enzyme affin-ity for F6P. The consequence of prod-uct activation is shown in the lower panel. As a function of time, product activation leads to an exponential rate of increase of enzyme activity. Most enzymes, in contrast, are subject to product inhibition, also shown in the lower panel (from Hochachka and So-mero, 1973).

situation which also facilitates the tighter control of PFK which is needed in muscle.

To complete the "cascade" activation of muscle glycolysis, the activities of PFK must be integrated with the next major "valve" in the pathway—pyruvate kinase. As the ADP formed in the PFK reac-tion is a substrate for the PK reaction, it serves as an important meta-bolic coupling mechanism between PFK and PK. In mammals, this ap-pears to represent the only adaptation available to the organism for integrating these two reaction steps. It is, however, a finely tuned mechanism. Muscle PK in mammals displays a five- to sixfold greater affinity for its substrates than does the liver isozyme; hence, it is able to compete for limiting substrate with high efficiency.

Finally, the pyruvate formed by the PK reaction is converted to lactate in the LDH reaction, regenerating NAD in the process (Figure 1).

In vertebrate muscle contracting under anoxic conditions, these events are so closely integrated that *lactate production is directly proportional to muscle work,* and the muscle sustains larger accumulations of lactate than ever occur in other tissues. Indeed, the final functional requirement for muscle glycolysis is some provision for the accumulation of huge quantities of lactate and for its subsequent metabolism.

The tolerance of high lactate levels

At least two specific mechanisms can be viewed as adaptations to the problem of huge lactate accumulations, and both of these appear to rely upon the elaboration of unique "kinds" of enzymes:

1. The LDH isozyme predominating in white muscle of vertebrates (designated the M_4 tetramer) is very insensitive to pyruvate inhibition, and this property allows large accumulations of lactate from pyruvate. In addition the M_4 LDH affinity for pyruvate is rather low compared to the H_4 (heart-type) isozyme. This property presumably prevents the M_4 enzyme from being readily saturated and hence prevents concentrations from rising without limit in muscle.

2. Perhaps the most important mechanism for dealing with high lactate accumulations is the development of a means for metabolizing the lactate produced. By far the largest portion of the lactate produced during anoxic muscle work in the vertebrates is delivered by the blood to glucogenic tissues (mainly the liver) where it is nearly quantitively converted to glucose or glycogen via the gluconeogenic pathway. In these terms, gluconeogenesis in the liver can be viewed as a mechanism for dealing with the lactate accumulation in muscle during anaerobic work. (It should be noted that gluconeogenesis plays other important intertissue functions, chief among these being the synthesis of glucose for the central nervous system).

Most of the reactions from lactate to glucose are catalyzed by enzymes of the glycolytic scheme and thus proceed by reversal of steps employed in glycolysis. However, there are four irreversible steps in the normal "downhill" glycolytic pathway which cannot be utilized in the "uphill" conversion of lactate (or pyruvate) to glucose. These are the steps catalyzed by PK, PGK, PFK, and HK. During glucose synthesis these reaction steps are bypassed by alternative reactions which are favourable in the glucogenic direction. Three such bypass reactions are now known, which in my frame of reference constitute an evolutionary "solution" to the problem of converting lactate and other triose precursors to glucose:

1. Pyruvate kinase is bypassed by the concerted action of at least two enzymes: pyruvate carboxylase catalyzes the carboxylation of pyruvate to form oxaloacetate (OXA), while PEP carboxykinase

catalyzes the decarboxylation of OXA to form PEP. Both reactions require a high-energy phosphate compound and their combined function, unlike the PK reaction, favours PEP formation from pyruvate. It should be noted parenthetically, that in liver the presence of these enzymes complicates the PEP branchpoint; therefore, it is not surprising to find that the liver PK displays catalytic and regulatory properties which are quite different from muscle PK (Llorente et al., 1970);

2. The PFK reaction is bypassed by FDPase which catalyzes the hydrolysis of FDP to F6P $+$ P_i;

3. The HK reaction is bypassed by G6Pase which catalyzes the hydrolysis of G6P to glucose $+$ P_i.

Both of the latter enzymes proceed in the glucogenic direction with a large free energy drop; they are therefore "ideal" bypass solutions to the problem of "uphill" reactions in lactate-to-glucose conversion.

Compensatory nature of anoxia adaptation in vertebrate muscle

From these considerations, we can conclude that anaerobic muscle work in the vertebrates is supported by three fundamentally adaptive characteristics:

1. A high glycolytic potential stemming from high concentration of glycolytic enzymes;

2. The occurrence of specific muscle isozymes kinetically attuned (a) for function in the glycolytic direction, and (b) for exponential rate of change from low-activity to high-activity states; and

3. A tolerance for huge accumulations of lactate with provision for its subsequent metabolism (and in particular for its subsequent reconversion to glucose).

These mechanisms in the vertebrates *compensate for the temporary absence of molecular oxygen in muscle by allowing for an impressive increase in anaerobic ATP-generating capacities. In the most skilled of air-breathing vertebrate divers, it is precisely these loci which we might expect to find even further adjusted.* Is this expectation in fact realized?

Enzyme adaptations to deep-diving in air-breathing vertebrates

At the moment we do not know the answer to the above question, for the biochemical problems associated with deep diving have not been examined in this context. Yet from the available data, it appears that the above compensatory adaptations reach their zenith in diving vertebrates such as aquatic turtles, marine mammals, and marine birds. Some of the most reliable information is to be found in studies of aquatic turtles, but the principles arising should be equally applicable to most deep-diving vertebrates.

On the basis of direct calorimetric measurements, Jackson (1968) has divided the diving period of turtles into three metabolic phases: During Phase I, the metabolic rate persists at the same rate as in the prediving condition, but oxygen tensions in the blood and the lungs rapidly fall. In Phase II, the metabolism falls precipitously and remaining oxygen reserves are exhausted. The remainder of the dive (Phase III), which can last for many hours (and in some species at low temperature, for many days!) is totally anoxic. All maintenance and work functions in Phase III are sustained by anaerobic metabolism, and the yield of energy is about 20 percent of that available to the organism in the prediving state. Apparently, anaerobic glycolysis accounts for all of the energy generated at this time. The unusual activity of this pathway in these species (probably reflecting very high levels of component enzymes) leads to unusually high lactate accumulations in the tissues and in the blood. Because even "aerobic-type" tissues, such as the heart, have a high glycolytic capacity, the usual kinetic differences between H_4 (heart-type) and M_4 (muscle type) LDHs are absent. In these organisms, as well as in diving ducks, penguins, and seals, the heart LDHs do not show the usual sensitivity to substrate inhibition (Altman and Robin, 1969; Markert and Masui, 1969; Blix and From, 1971). These LDH activities maintain pyruvate levels at 0.1mM, while lactate levels can rise to over 60 mM !

Belkin (1963) first provided unequivocal demonstration of the critical importance of anaerobic glycolysis to the anoxic turtle. He selected a metabolic poison, iodoacetate, whose chief locus of action is known to be the triose phosphate dehydrogenase reaction in glycolysis. If anoxic survival depended upon glycolysis, inhibition of this enzyme by iodoacetate injection should lead to a greatly reduced tolerance to anoxia. This prediction was verified. Jackson (1968) then observed by direct calorimetry that iodoacetate injection leads to a predicted drop in the remaining energy metabolism of the anoxic organism. These studies underline the central role of glycolysis in the adaptation to prolonged diving in turtles, and presumably a similar situation occurs in diving marine mammals and birds. By this mechanism enough energy is generated to sustain turtles in dives of many hours duration; yet sooner or later even the diving turtle surfaces and "repays" its oxygen debt, for here as in most other vertebrates, the biochemical strategy of anoxia adaptation depends upon an ultimate return to aerobiosis. It is a strategy that merely compensates for the temporary absence of molecular oxygen in the body, and hence it is a poor strategy for the exploitation of oxygen-free environments on a sustained basis.

Environmental limitations on oxygen availability

In contrast to the vertebrates, massive exploitation of oxygen-free zones has been achieved by many invertebrate organisms. By and large, the oxygen-free zones in the marine environment that I am thinking of are the benthic sediments of the open oceans and of inshore regions. Organisms which burrow in these benthic sediments apparently are highly tolerant of anoxia as an adaptation to the near absence of molecular oxygen in the interstitial waters (see Mangum, 1970). In many instances, such sediments are in a highly reduced state (Fenchel and Riedl, 1970) and organisms such as bivalve molluscs, turbellaria, nematodes, or annelids may have been forced to "adapt" to what I am terming an "environmental limitation on oxygen availability" (to contrast it with the physiological limitations in vertebrate divers). Many gill-breathing and essentially sessile invertebrates encounter similar problems in the intertidal regions: although exposed at high tide to the infinite abundance of atmospheric oxygen, their modes of respiration (and life) make it quite unavailable to them. Superficially, there may be some resemblance between the air-breathing vertebrate in water and the gill-breathing invertebrate in air, but because organisms such as intertidal bivalves are essentially non-motile, they cannot return to the sea to pay off their oxygen debt whenever necessary. Hence, the biochemical strategies of adaptation they have utilized are very different from those described above for the vertebrates.

Anaerobic metabolism in facultative anaerobic invertebrates

The best available data on this problem come from studies of intertidal bivalve molluscs, but the basic principles emerging may be instructive in considering anoxia adaptation in other invertebrates. In bivalve molluscs such as *Crasssostrea*, *Mytilus*, and others, glycogen is an important (but is not the sole) source of carbon and energy during anoxic muscle work (Hochachka and Mustafa, 1972). Many phases of glycogen mobilization in these species are similar to those in vertebrates, but at the level of PEP, the metabolic pathways which are functional depend upon the availability of oxygen. In the presence of oxygen,

$$PEP \rightarrow pyruvate \rightarrow \rightarrow CO_2 + H_2O$$

In the absence of oxygen, a second pathway is activated:

$$PEP \rightarrow OXA \rightarrow malate \rightarrow fumarate \rightarrow succinate$$
$$\searrow pyruvate \rightarrow alanine$$

Thus, in these organisms, alanine and succinate accumulate as end products of anaerobiosis. The two enzymes strategically situated at the

PEP branchpoint, pyruvate kinase (PK) and PEP carboxykinase (PEPCK), are apparently adapted for *either/or, reciprocal* function. That is, either PK is active during aerobiosis or PEPCK is active during anoxia, but the two do not appear to be able to function simultaneously in optimal manners (Hochachka and Mustafa, 1972). The metabolic signals involved in the anaerobic-aerobic transition at this point in metabolism are summarized briefly in Figure 3. The unique metabolic events in these organisms are so closely integrated that during anoxia, *muscle contractile work is apparently proportional to the production of succinate (and alanine)*; lactate, in contrast, usually is not produced in high quantities because of the "deletion" of the LDH enzyme.

Metabolic coupling of additional substrate-level phosphorylations to glycolysis

Insight into the functional significance of this unique metabolic organization arose from the realization that the fate of pyruvate formed during anoxia is transamination to alanine:

pyruvate $+$ glutamate $\rightarrow$ alanine $+$ α-ketoglutarate.

The alanine accumulates as a second end product of anaerobiosis, but the subsequent metabolic fate of α-ketoglutarate (α-KGA) is of greater interest (Figure 4). Firstly it is converted to succinylCoA by the action of α-KGA dehydrogenase:

α-KGA $+$ CoASH $+$ NAD $\rightarrow$ succinylCoA $+$ CO_2 $+$ NADH

Secondly, the α-KGA dehydrogenase reaction sets the stage for the conversion of the thiolester bond energy of succinylCoA into nucleoside triphosphate. The reaction, catalyzed by succinic thiokinase, is exergonic and can utilize either GDP or IDP as cosubstrate with succinylCoA, generating GTP or ITP. This energy-yielding reaction, an integral "substrate-level phosphorylation" step in the Krebs cycle, is utilized as an anaerobic mechanism for supplanting aerobic metabolism in certain mammalian tissues (Cohen, 1968) and presumably serves an analogous function in facultative anaerobic invertebrates. In both systems, for every two moles of ATP generated by glycolysis, a third mole can be generated at the succinic thiokinase step. In both systems, however, some provision must be made for the regeneration of NAD required for α-kGA dehydrogenase activity. In facultative anaerobes the most likely candidate for the job is fumarate reductase, which couples the oxidation of NADH with the reduction of fumarate to succinate (Hammen, 1969). It is this function which explains the unique kinetic features of the enzyme (a relatively high affinity for fumarate and a low affinity for succinate) as well as its high activity in these organisms (Saz, 1971). In addition, fumarate reductase is prop-

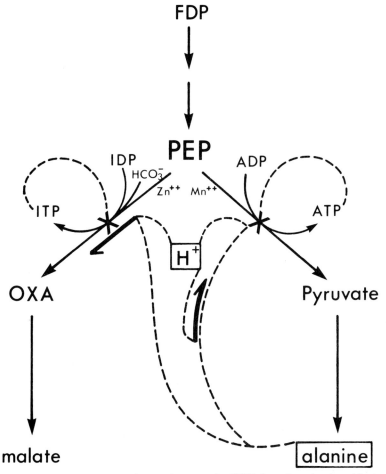

Figure 3. Known control interactions at the PEP branch point in adductor muscle of the oyster. Effective activation or deinhibition are shown with dark arrows; effective inhibition, by a dark cross. In addition, FDP is an established feed-forward activator of pyruvate kinase, and this mechanism could account for some PK activity during anoxia (from Hochachka and Mustafa, 1972).

erly positioned in the mitochondrion for the delivery of NAD to the α-KGA dehydrogenase reaction (Saz, 1971; Hammen, 1969).

From these considerations, one can view the unique pattern of anaerobic energy metabolism in invertebrate facultative anerobes as a means for coupling an additional substrate-level phosphorylation to glycolysis, thereby increasing the overall potential yield of high-energy

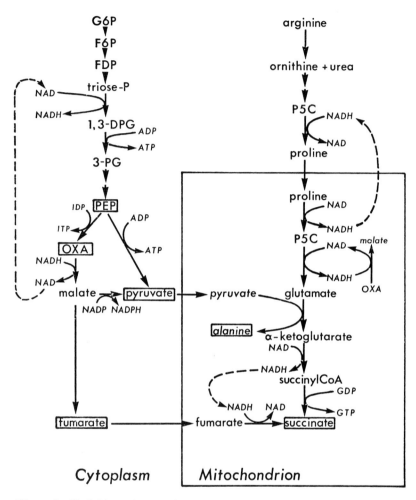

Figure 4. Probable pathways of anaerobic intermediary metabolism in inverte-brate facultative anaerobes. (data from Hochachka and Mustafa, 1972). The key role of succinic thiokinase in this metabolic organization was an important in-sight into this problem for it led to the realization that many of the known end products of anaerobiosis in these animals could be produced from their CoA derivatives. In addition to succinate (shown above), propionate, acetate, methyl-butyrate, isobutyrate, and isovalerate are known end products whose formation can proceed according to the reaction scheme:

acyl CoA + ADP + P_i → ATP + acid end product.

This insight explains the functional "advantage" of producing the above end products compared to producing lactate. The acyl CoA derivates are thought to phosphate compounds. Stoichiometric coupling can be envisaged at two sites:

1. At the pyruvate-glutamate transaminase reaction, in which one substrate derives from glycogen while the second derives from the amino acid pool; and

2. At the fumarate reductase step, in which the substrate (fumarate) derives from glycogen, while the reducing equivalents (in the form of NADH) derive from α-KGA.

Hence, these two coupling reactions can be viewed as a means for achieving the *simultaneous mobilization of two energy reserves—carbohydrate and amino acids—during anoxic excursions.*

Exploitive nature of anoxia adaptation in invertebrate facultative anaerobes

In contrast to the vertebrate situation, then, three processes seem to have been favoured during biochemical adaptation to limited oxygen availability in the invertebrates:

1. The deletion of certain enzymes, in particular, the deletion of LDH to avoid metabolic *cul de sacs,* such as lactate production;

2. The modification of the kinetic properties of certain key branchpoint enzymes (PK and PEPCK) to allow an efficient transition from aerobic to anaerobic metabolism; and

3. The coupling of other "substrate-level" phosphorylations to those occurring in glycolysis, thus increasing the potential yield of high-energy phosphate compounds.

These molecular adaptations clearly are of an "exploitive" nature, in the sense that they apparently facilitate the organism's utilization of oxygen-free environments on a sustained rather than on a temporary basis. At the moment, it is difficult to assess just how frequently this exploitive strategy is utilized by marine organisms. Most of the cur-

---

arise by a two-step amino acid fermentation:

amino acids $\rightarrow$ 2-ketocarboxylates $\rightarrow$ acyl CoA derivatives $+$ CO_2

$$\overset{\frown}{NAD \quad NADH}$$

The first step is catalyzed by various transaminases; the second by ketocarboxylate dehydrogenases. The conversion of glutamate to succinate is but one example of such a round of reactions. However, it clearly indicates another important principle of anoxia adaptation: *each ketocarboxylate dehydrogenase (e.g., α-KGA dehydrogenase) must be functionally linked in a 1:1 activity ratio to one thiokinase (e.g., succinic thiokinase) so that the system neither depletes nor accumulates reduced Coenzyme A.* Each such functional unit of enzymes is termed a Coenzyme A cycle, or a CoA cycle for short, because in regard to CoA reserves, the system is cyclic and catalytic. See Hochachka et al., Amer. Zool. 13, 543-555 (1973) for a fuller treatment of this area. Chart from Hochachka and Mustafa, 1972.

rently available information comes from studies of intertidal organisms. However, comparable physiological studies of benthic invertebrates are already appearing (Mangum, 1970), and it is probable that the underlying biochemical mechanisms are similar to those noted above. Be that as it may, the kinds of macromolecular functions described above come under the direct influence of two other important environmental parameters: the external temperature and pressure. How have marine organisms adapted to the direct impingement of these physical parameters upon their cellular chemistry? To analyze this problem, I will briefly consider the basic effects of temperature and pressure upon enzymes from heterothermic organisms in general before considering what is known of enzyme mechanisms by which marine organisms adapt to the two factors.

Adaptations to Temperature

Effects of temperature on catalytic and regulatory properties of enzymes

Classically, the effect of temperature on enzyme reaction rates was interpreted according to the Arrhenius relationship given by:

$$\mu = 2.3 \, R \, \frac{\log K_{T_1} - \log K_{T_0}}{\dfrac{1}{T_0} - \dfrac{1}{T_1}}$$

where μ is the activation energy, R the gas constant, K the rate constant for the reaction at two different temperatures, T_1 and T_0, given in degrees Kelvin. A typical graphical presentation of this equation is shown in Figure 5 (for curve labeled, V_{max}). For workers interested in temperature adaptation, two portions of the Arrhenius curve are of interest—the activation energy, which is directly proportional to the slope of the line, and the thermal optimum. Earlier workers considered that either characteristic or both should correlate with the habitat temperature, but as more data became available, it became evident that neither does (see Hochachka and Somero, 1971). In regard to thermal optima, the argument is simply that for most enzymes these are far beyond the lethal limits for the organism. The situation is somewhat less clear regarding the value of μ. Basically, the idea commonly held is that selection would favor enzymes of high catalytic efficiency (and hence low μ value) in organisms living at low temperatures. Where μ is an acceptable measure of catalytic efficiency, the correlation may hold.

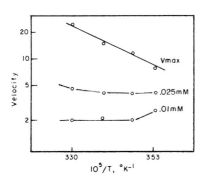

Figure 5. Arrhenius plot of liver citrate synthase activities from cold (2°C) acclimated trout at saturating oxaloacetate and acetylCoA concentrations is shown in the curve labeled Vmax. The calculated activation energy is 8.8 kcal/mole, corresponding to a Q_{10} of about 1.7. In the two lower curves, oxaloacetate concentrations were at saturating levels, but acetyl-CoA levels were 0.01 and 0.025 mM as shown. It is evident that the Q_{10} is dramatically decreased at low substrate concentrations. (Data from Hochachka, 1971. Similar data for several different enzyme systems are available in Hochachka and Somero, 1971.)

But the turnover number of an enzyme can also be strongly influenced by activation entropy, which has been largely ignored by earlier studies. This may account for the absence of an obvious correlation between the μ value for any two homologous enzymes and the thermal habitat of the parent species, whereas such a correlation does appear to exist for the turnover number of enzymes. (See Hochachka and Somero, 1973, for a further discussion of this point.)

The situation, however, is still more complex. As is evident in Figure 5, the slope of Arrhenius plots can vary dramatically depending upon substrate concentrations. At low substrate levels, the temperature coefficient (Q_{10}) is dramatically reduced and can take on values of less than 1.0. The explanation for such seemingly anomalous thermal properties is in the nature of the catalyst. For a large number of enzyme systems, we have found that the apparent Michaelis constant (K_m) varies directly with temperature over at least a part of the biological range. Over this range, as temperature increases, the apparent enzyme-substrate affinity decreases in such a way as to compensate for the thermal energy change (Figure 6). Thus, under low (physiological) substrate concentrations, *the reaction rate is being determined by the kinetic properties of the catalyst rather than thermodynamic parameters* (Hochachka and Somero, 1971).

In many ways, there is a strong similarity between the effects of temperature and the effects of metabolite modulators on enzyme-substrate affinities. A particularly clear example is to be found in the case of PFK from the arctic king crab, where the effects of low temperature and of the positive modulator, AMP, are strikingly comparable

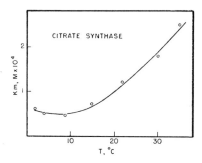

Figure 6. Effect of temperature on the K_m of acetylCoA for liver citrate synthase from cold acclimated rainbow trout (Hochachka, 1971). As temperature increases, the apparent enzyme-substrate affinity (reciprocal of the K_m) decreases, thus compensating the reaction for changes in the thermal energy of the reactants.

(Freed, 1971) (Figure 7). In this example, as in the case of many regulatory enzymes, catalytic activity is regulated by the binding of specific metabolites at sites separate from the catalytic site; usually, these modulators alter the apparent enzyme-substrate affinity with no necessary effect on the maximal velocity (V_{max}) of the reaction. Positive modulators decrease the apparent K_m, and sometimes convert the sigmoidal curve into a hyperbolic one. (At low substrate levels, this is tantamount to activating the enzyme.) In both these effects, the action of positive modulators is analogous to that of low temperature. Somero (1969) also has described an elegant example of this in the case of king crab pyruvate kinase: At low temperatures, the apparent K_m of the substrate (PEP) is reduced, and substrate saturation curves become hyperbolic rather than sigmoidal.

Another fundamental characteristic of regulatory enzymes is that saturation curves for positive (or negative) modulators are often

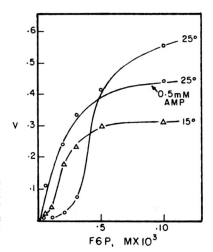

Figure 7. Comparison of the effects of the positive modulator, AMP, and of low temperature upon the activity of king crab phosphofructokinase (Data from Freed, 1971; redrawn by Hochachka, 1971).

sigmoidal. An important implication of this is that small changes in modulator concentration can lead to relatively large changes in the activity of the enzyme. This property coupled with others, such as product activation, leads to regulatory behavior which approximates an "on-off" switch. In general, "on-off" switch mechanisms of poikilothermic enzymes seem to be less temperature sensitive than are enzyme-substrate interactions (see Hochachka and Somero, 1971). This admirably suits poikilothermic existence, for the regulation of enzyme activity can then be achieved with equal efficiency over the entire biological temperature range, *irrespective of the effects of temperature on maximum velocities.*

Because of the fundamental role of K_m modulation in enzyme regulation in general, then, it is not too surprising that the effects of change in the thermal energy of reactants can be counteracted by other appropriate and instantaneous effects of temperature, presumably upon enzyme conformation and hence upon enzyme-substrate affinities. To what extent are these mechanisms operative in enzymic adaptation to thermal profiles of the sea? What, in other words, are the strategies of enzyme adaptation to temperature utilized by organisms in different positions in the water column?

Factors determining choice of thermal adaptational strategy

In considering the above questions, it is important to recall that different ectotherms range in their thermal tolerances (and requirements) from narrow stenothermality (abyssal and benthic species) to relatively broad eurythermality (vertical migrants and many surface species). We would therefore expect different strategies of enzyme adaptation to be utilized by these organisms, depending upon their position in the water column. Thus, most abyssal organisms are probably stenothermal, tightly adapted to a very narrow thermal range. Enzyme function in these organisms may likewise be tightly linked to a narrow thermal regime. Unfortunately, very little information is available on enzymes of abyssal organisms, and those studies which have been done (see Mustafa et al., 1971; Moon et al., 1971) have concentrated on pressure adaptation problems. Recent studies of liver FDPase from an abyssal fish (*Coryphaenoides*) include a comparison of enzyme-substrate affinities at different temperatures. For this enzyme, enzyme-substrate affinities are maximal at near 0°C (Hochachka et al., 1970), a result that is not surprising in view of previous enzyme studies of Antarctic fishes, which are also highly stenothermal (see Hochachka and Somero, 1971).

In the case of eurythermal surface species such as the salmonids, there is a substantial ability to maintain key enzyme-ligand interactions relatively independent of temperature. In comparison, very few eurythermal vertical migrants have been studied in this context. However, Mustafa (unpublished data) has obtained excellent data on PK from a vertically migrating squid species. For this enzyme, the apparent K_m (PEP) is thermally insensitive over a broad thermal range (covering that normally encountered in its habitat), but increases sharply at the upper limits of the biological thermal range.

Similar effects are observed for the enzyme from intertidal bivalves (Figure 8), which are also extremely eurythermal. In both cases, selective forces favoring a temperature-independent K_m appear to outweigh the compensating advantages of a K_m directly dependent upon temperature. This is a strategy that in effect designs an enzyme's regulatory properties in such a way as to maintain control of its activity independent of temperature—an eminently "sensible" strategy for a mainline catabolic enzyme, such as PK, which must perform its catalytic and regulatory functions in an organism whose body temperature may be changing by over $10°$ C during daily vertical migrations in the water column.

Interestingly enough, enzymes of midwater fishes seem to display lower activation energies than do homologous enzymes from either surface or abyssal species. Thus, the activation energy for PK of a midwater sea bass (*Ectreposebastes*) is only about half that ob-

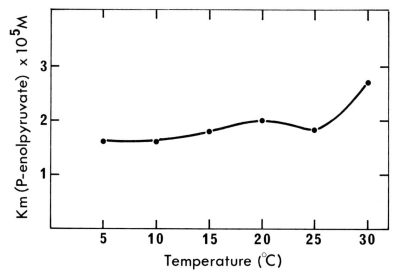

Figure 8. K_m-temperature relationship for pyruvate kinase of adductor muscle of the oyster, *Crassostrea gigas* (from Mustafa, unpublished data).

served for the homologous enzyme in trout, the leatherjacket, or the rattail (Moon et al., 1971; Mustafa et al., 1971). Similarly, the liver enzyme, FDPase, extracted from the midwater fish, *Stomias*, displays an unusually low activation energy when compared to the homologous enzyme in either surface or abyssal fishes (Hochachka et al., 1970). All else being equal, these low activation energies indicate *enzymes of very high catalytic efficiency* (i.e., high turnover number) and of reduced temperature-sensitivity even under saturating concentrations of substrate.

If these results are typical for enzymes of midwater organisms, they may supply a sufficient mechanism for (1) maintenance of regulatory functions in the face of changing external temperature, and (2) maintenance of low (and controlled) Q_{10} values through biological temperature ranges. However, a plethora of questions concerning the role of enzymes in temperature adaptation of midwater organisms remain: What are the effects of temperature on enzyme-modulator interactions? Do these organisms possess unusually small (or large) numbers of isozymes for any given enzyme? Do these isozymes play an additional role in thermal adaptation? And, most significantly, what are the interacting effects of changing temperature and pressure upon catalysis and control of catalysis?

Adaptations to Pressure

Pressure as an environmental variable

Hydrostatic pressure impinges directly and unavoidably on the cellular chemistry of all organisms. But even though it is an environmental parameter with which all forms of life must cope, biologists largely have tended to ignore it in their studies of environmental physiology and biochemistry. In part this neglect is well justified, for the organisms which biologists favor in their studies almost exclusively inhabit the terrestrial environment or the first few meters of the waters of the earth. These organisms encounter only modest absolute pressures and essentially no variation in pressure: from sea level to the highest point on earth, atmospheric pressure varies only about fourfold, from 1 to 0.25 atmosphere. Changes in pressure over this range have little effect on the chemistry of the cell. Only in the marine environment and in deep freshwater lakes does pressure assume importance as an environmental parameter extensively affecting biological systems. Sea water is nearly 1,000 times as dense as air and hydrostatic pressure doubles with each 10 meters of depth. The average abyssal pressures of the ocean floor range between 300 to 500 atmospheres; in the deepest oceanic trenches, pressures exceed 1,000 atmospheres.

In addition to encountering severe absolute pressures, marine organisms also may be subjected to wide variations in pressure, either diurnally or at different times throughout the life cycle. For example, fishes such as the Myctophids undergo daily 300 to 500 meter vertical migrations in the water column, depths corresponding to pressure changes of some 30 to 50 atmospheres. Other midwater organisms (squids, other fishes such as *Stomias* and *Ectreposebastes*) undergo vertical migrations which are two- to fivefold greater. Many benthic organisms appear to have pelagic larval stages, and through a complete life cycle these species can encounter even greater pressure ranges. As we have already seen, organisms migrating through the water column also face changes in temperature, often of 10-15° C range. Thus, we find that pressure stress is commonly associated with temperature stress, and as a result the biochemical machinery of many marine organisms must be adapted to deal with them simu'taneously. I have already discussed potential enzyme mechanisms in temperature adaptation. Are similar mechanisms operative in pressure adaptation? Before examining this problem in detail, let us briefly review the basic effects of pressure on chemical reaction rates.

Fundamental theory of the action of pressure on enzyme reaction rates

In analyzing the factors instrumental in establishing the pressure sensitivity of an enzyme reaction, it is essential to distinguish carefully between parameters which govern the final equilibrium of the reaction and those which determine how rapidly this equilibrium is attained. In regard to temperature, the equilibrium which the reaction attains is set by the free energy change ($\triangle G°$) which occurs as substrates are converted to products; the *rate* at which equilibrium is attained is governed by a second free energy function, the free energy of activation.

We find analogous parameters involved in the kinetics and thermodynamics of pressure effects. As we might expect, the key parameters which establish the direction and magnitude of pressure effects on a reaction involve volume changes. If the volume of the system containing the reactants is greater than the volume of the system containing the products, then pressure will shift the equilibrium toward product formation. If the volume of the reactants is less than the volume of the system containing the products, pressure will favor the accumulation of reactants at equilibrium. Formally, these relationships can be described as follows:

$$\Delta V = 2.3\ RT\ \frac{\log_{10}K_{P_1} - \log_{10}K_{P_2}}{P_1 - P_2}$$

where ΔV is the volume change of the reaction, R the gas constant with the value 82.07 cm³/mole, T the absolute temperature in degrees Kelvin, and K the velocity constant at pressures P_1 and P_2 atmospheres.

The overall volume change of the reaction, ΔV, has its analogue in the overall free energy change of the reaction; *both parameters indicate the final equilibrium, but say nothing about the speed with which this equilibrium is attained.*

The critical factor which determines the effect of pressure on reaction rate, as opposed to reaction equilibrium, is the volume change which accompanies the formation of the transitory activated complex. We can equate this volume change of activation ($\Delta V^{\ddagger}$) with the volume differences between the systems containing nonactivated reactants and activated reactants.

The equation describing the dependence of velocity on pressure is formally analogous to the Arrhenius equation relating reaction rates to temperature:

$$\Delta V^{\ddagger} = 2.3\ RT\ \frac{\log_{10}K_{P_1} - \log_{10}K_{P_2}}{P_1 - P_2}$$

When the volume of the activated complex exceeds the average volume of its constituents outside the complex, pressure inhibits the reaction. If the volumes become equal, there is no change in the reaction rate under pressure. When the volume of the activated complex is less than that of the reactants, pressure increases the reaction rate. In this latter case, for enzyme-catalyzed reactions, the reaction rate can go through a pressure optimum. Either one of two factors usually determines such an optimum:

1. The reaction can become diffusion limited due to large increases in the vicosity of the medium, or
2. High pressures can lead to enzyme denaturation.

As both factors can be influenced by temperature, the actual pressures at which the optimum occurs often vary with temperature.

The sign of $\Delta V^{\ddagger}$ and its metabolic consequences

With regard to pressure, biochemical reactions can be classified into three categories:

1. Pressure-activated reactions, with a negative $\triangle V^{\ddagger}$,
2. Pressure-inhibited reactions, with a positive $\triangle V^{\ddagger}$, and
3. Pressure-independent reactions proceeding with no net volume change.

The chief determinants of the category of response observed are the structures of the reactants and of the transition "activated" complex. Since these can vary with temperature in enzyme-catalyzed reactions, it follows that pressure can bring about all three effects on a given enzyme, depending upon the temperature.

The combined consequences of (1) differential pressure effects among different enzymes, and (2) differential pressure responses by a single enzyme at different temperatures, render metabolism exceedingly "vulnerable' 'to pressure changes. This is particularly important where two or more enzymes compete for a single, common metabolite, for such differential effects of pressure then profoundly alter the flow of carbon through metabolic branching points and thus strongly "unbalance" the partitioning of carbon and energy between various cellular processes. Indeed, these differential effects of pressure on biological processes may represent the most essential biochemical problem in pressure adaptation.

The magnitude of $\triangle V^{\ddagger}$ for enzyme catalyzed reactions

Whereas the sign of $\triangle V^{\ddagger}$ determines whether the pressure effect is a stimulation or a retardation in rate, the absolute value of $\triangle V^{\ddagger}$ will determine how sharply pressure affects reaction velocity. Enzymic reactions characteristically exhibit much larger volume changes during activation than nonenzyme-catalyzed reactions. No doubt this difference is due to volume changes occurring during catalysis and involving the enzyme protein itself.

For non-catalyzed reactions, $\triangle V^{\ddagger}$ values normally are less than 20 cm^3/mole. $\triangle V^{\ddagger}$ values for enzymic reactions may surpass 100 cm^3/mole. $\triangle V^{\ddagger}$ values of this magnitude would lead to approximately tenfold changes in reaction rate as pressure is increased from 1 atmosphere to 1,000 atmospheres, assuming that $\triangle V^{\ddagger}$ does not vary with pressure. In theory, therefore, reaction rates occurring in deep sea organisms could be altered three- to tenfold by pressures of 300 to 1,000 atmospheres.

These effects of pressure are large—large enough to suggest that any abyssal or vertically migrating organism capable of reducing or minimizing them would gain a strong selective advantage. The wide abundance of mesopelagic and abyssal organisms indicates that the effects of pressure on enzyme reaction rates are either accommodated or circumvented. The question is how.

Strategies of enzyme adaptation to pressure in marine organisms

——*Factors determining choice of strategy.* As in the various modes of adaptation to other environmental parameters, the strategy of adaptation to pressure at the enzymic level appears to depend upon the time available for the adaptation. Vertical migrating midwater organisms, for example, face pressure changes on a daily basis. Here, the adaptive strategy must take into account both the *absolute pressure* and *its rate of change.* In contrast, abyssal, benthic organisms do not encounter serious change in pressure, and the main problem is exposure to continuously high pressure. It is evident, therefore, that the selective forces impinging on cellular metabolism in marine organisms will depend strongly on their positions in the water column and on their life style. Recent studies of the effects of pressure upon enzymes extracted from surface, midwater, and abyssal fishes have shed some insight into the kinds of evolutionary adjustments which have occurred. The catalytic and regulatory adaptations of several enzymes were examined, including pressure-activated. pressure-inhibited, and pressure-independent ones.

Position and function of FDPase in metabolism. Fructose diphosphatase (FDPase) in the presence of Mg^{2+} or Mn^{2+} catalyzes the hydrolysis of FDP to yield F6P and P_i. In vertebrates, the enzyme is elaborated in highest quantities in those tissues (liver and kidney) which can synthesize glucose from precursors such as amino acids and 4-carbon keto acids. In these tissues, FDPase is an important control site in the gluconeogenetic pathway since concurrent function of both FDPase (FDP → F6P + P_i) and of phosphofructokinase (F6P + ATP → FDP + ADP) within a single cell compartment would lead to a futile cycling of carbon and a net hydrolysis of ATP. It is widely accepted that AMP serves as an important "coupling" modulator at this point in metabolism; under energy-depleted conditions, high AMP levels inhibit FDPase and simultaneously activate PFK. Most of the available information on pressure effects compares the enzyme from the trout, a surface-dwelling species, to the rattail, an abyssal benthic species (Hochachka et al., 1971 a,b).

In these comparative studies of FDPase, certain similarities in pressure responses were observed. Specifically:

1. *pH profile effects.* The pH dependencies of both enzymes are affected similarly by increasing hydrostatic pressure (Figure 9). At alkaline pH values (pH 7.5 and greater), the V_{max} characteristics of both the rattail and trout enzymes are strongly accelerated by increased pressure.

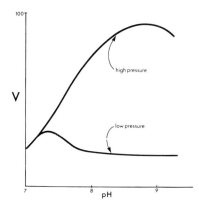

Figure 9. Effect of pH on the pressure responses of FDPases. The pattern observed is similar for the FDPases extracted from abyssal and surface organisms (from Hochachka and Somero, 1973).

2. *Activation energy effects.* For both enzymes, increases in pressure at neutral pH increase the activation energy of the reaction. At alkaline pH values this effect is reversed.

3. *Reduced pressure sensitivity at high temperatures.* For the trout and rattail enzymes, increases in temperature reduced the effects of pressure on enzymic activity. This observation is commonly made in pressure studies. The basis of this effect may lie in the roles of hydrophobic interactions in low temperature and high pressure denaturation phenomena.

4. *Pressure sensitivity of the native enzyme.* A further similarity between the surface and abyssal enzymes is their extreme pressure liability under conditions where substrate and cofactor are absent. Again, the instability of "naked" enzymes is a frequent observation in studies of pressure and temperature effects.

I list these similarities between two FDPase variants adapted to function at widely different pressures not to suggest that the enzymes are functionally identical, but rather to indicate that certain characteristics of an enzymic reaction may be inseparable features of a basic chemical transformation. For example, the free energy change which occurs as substrate is converted to product is of similar magnitude and same sign for all variants of an enzymic reaction. Likewise, the steric changes which must occur in the substrate during the activation may be identical for all variants of a reaction. That is, certain basic thermodynamic and steric parameters are necessary concomitants of any given chemical transformation. These parameters are fixed for a given enzymic reaction, regardless of the species or tissue in which the reaction occurs. Selection cannot tailor these parameters in any major way to better adapt the enzyme for function in its particular milieu.

If we can make any generalization about these relatively "inviolate" parameters, it is that they are involved in establishing the basic catalytic capacity of the enzyme. *In no case do these parameters play significant regulatory roles.* In contrast, an enzyme's regulatory functions, which determine the extent to which the enzyme's catalytic potential is actually utilized, are not fixed characters, identical in all variants of that enzyme. Rather, it is the regulatory properties of enzymes which in essence define the variant nature of enzymes, at least in a functional sense. These in all cases involve enzyme-ligand interactions: enzyme-substrate, enzyme-cofactor, enzyme-modulator interactions. What, then, is the influence of pressure on these all-important regulatory affinities between enzymes and lower molecular weight ligands? I can illustrate their role in enzyme adaptation to pressure by considering the effect of pressure on interactions between FDPase and its substrate, FDP.

FDPase-FDP interactions. Although the V_{max} for the FDPase reactions of trout and rattail is accelerated by pressure, FDPase-FDP interactions are affected by pressure quite differently in the two systems (Figure 10). For trout FDPase, the activity of the enzyme at physiological substrate concentrations is reduced at high pressure due to a sharp rise in the apparent K_m of FDP as pressure increases. For the abyssal enzyme a small increase in the apparent K_m of FDP occurs as pressure increases. However, because this effect is slight, the joint effects of the K_m change and the turnover number change are balanced, so that the rates of FDPase activity at physiological substrate concentration are pressure-independent (Figure 10). Thus the acceleration of catalysis *per se,* as evidenced by the rise in V_{max} with increasing pressure, is just sufficient to balance the slight decrease in activity at physiological substrate concentrations which stems from a decrease in FDPase-FDP affinity. In consequence, FDPase activity is independent of pressure in the abyssal fish.

This example illustrates a key point of enzymic adaptation: the physiologically important attributes of an enzyme, in this case the rate of the reaction at physiological substrate concentrations, are often the *net result of opposing changes in the total catalytic capacity, on the one hand, and the regulatory use of this catalytic capacity on the other.* For temperature effects, we have already seen examples whereby the influence of a temperature change on the V_{max} of a reaction is largely or fully counteracted by an opposing change in enzyme-substrate affinity. We now see that pressure effects, operating at the loci of turnover number and enzyme-substrate affinity, lead to a similar homeostasis of function. In temperature and pressure adaptations, therefore, it appears that a fixed, inviolate property of an enzyme—specifically, the energy changes involved in activation in the case of temperature and the

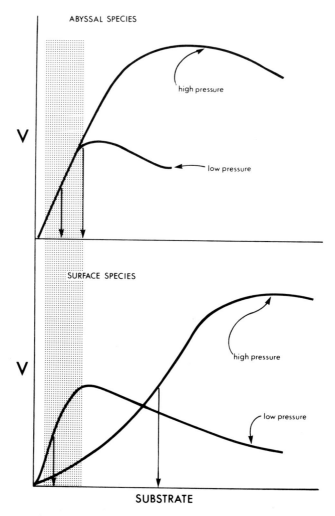

Figure 10. Substrate saturation curves for liver FDPases from surface and abyssal fishes. Approximate physiological range of FDP levels is shown by shading. In the surface fishes, pressure has a large effect on the K_m substrate; reaction rates are reduced at high pressures because of this reduced affinity for FDP. When the enzyme is saturated, high pressure activates the reaction velocity (V). In the abyssal species, the K_m still increases with pressure but only to the extent that it counteracts the pressure activation of catalysis; in consequence, at K_m values of substrate, the reaction velocity is independent of pressure. The effects of pressure on cofactor (Mg^{++}) saturation of these two FDPases are essentially identical to those for FDP saturation shown in a stylistic manner above (from Hochachka and Somero, 1973).

volume changes of activation in the case of pressure—of necessity perturbs enzymic function as the temperature or pressure of the environment changes. However, the organism compensates for the potentially adverse effects of this change by varying another enzymic parameter which is not a fixed attribute of the particular chemical transformation.

If these effects are general, they supply sufficient mechanism for avoiding drastic effects of high or varying pressure upon the cell chemistry of marine organisms. But, as in the case of temperature adaptation, in pressure adaptation there are a multitude of interesting problems awaiting exploration. Among these unanswered questions the following appear particularly relevant:

1. Do marine organisms encountering large pressure changes at different stages in their life cycles acclimate to pressure, much as ectothermic species acclimate to temperature?

2. Are the higher-order structures of enzymes from deep sea organisms pressure-labile like those of terrestrial organisms?

3. Are the functions of transcription and translation pressure sensitive?

4. Are the structures and functions of membrane-bound proteins unusually pressure sensitive in pressure-adapted organisms? Have membrane proteins in deep sea organisms become barophyllic? Are the phospholipid components of membranes varied as a function of the depth at which the organism lives?

5. Lastly, what biochemical mechanisms of pressure adaptation occur in deep-diving homeotherms such as whales, seals, and sea birds? Are there unique problems associated with the tolerance of large-scale pressure changes in a system held at a high and constant temperature?

The search for answers to these questions should raise new challenges to the science of marine biology.

Literature Cited

Altman, M., and E. D. Robin. 1969. Survival during prolonged anaerobiosis as a function of an unusual adaptation involving lactate dehydrogenase subunits. Comp. Biochem. Physiol., *30:* 1179-1187.

Blix, A. S., and S. H. From. 1971. Lactate dehydrogenase in diving animals—a comparative study with special reference to the eider. Comp. Biochem. Physiol., *40B:* 579-584.

Belkin, D. A. 1963. Anoxia: tolerance in reptiles. Science, *139:* 492-493.

Cohen, J. J. 1968. Renal gaseous and substrate metabolism *in vivo:* relationships to renal function. Intl. Congress Physiol. Sci., *6:* 233-234.

Drummond, G. I. 1971. Microenvironment and enzyme function: control of energy metabolism during muscle work. Amer. Zoologist, *11:* 83-97.

Fenchel, T. M., and R. J. Riedl. 1970. The sulfide system: a new biotic community underneath the oxidized layer of marine sand bottoms. Marine Biol., *7:* 255-268.

Freed, J. M. 1971. Temperature effects on muscle phosphofructokinase of the Alaskan king crab, *Paralithodes camtschatica*. Comp. Biochem. Physiol., *39B*: 765-774.

Hammen, C. S. 1969. Metabolism of the oyster, *Crassostrea virginica*. Amer. Zoologist, *9*: 309-318.

Hochachka, P. W. 1971. Enzyme mechanisms in temperature and pressure adaptation of off-shore benthic organisms: the basic problem. Amer. Zoologist, *11*: 425-435.

Hochachka, P. W., D. E. Schneider, and A. Kuznetsov. 1970. Interacting pressure and temperature effects on enzymes of marine poikilotherms: catalytic and regulatory properties of FDPase from deep and shallow-water fishes. Marine Biol., *4*: 285-293.

Hochachka, P. W., H. W. Behrisch, and F. Marcus. 1971. Pressure effects on catalysis and control of catalysis by liver FDPase from an off-shore benthic fish. Amer. Zoologist, *11*: 437-449.

Hochachka, P. W., D. E. Schneider, and T. W. Moon. 1971. The adaptation of enzymes to pressure. I. A comparison of trout liver FDPase with the homologous enzyme from an off-shore benthic fish. Amer. Zoologist, *11*: 479-490.

Hochachka, P. W., and G. N. Somero. 1971. Biochemical adaptation to the environment. In *Fish Physiology*, Vol. 6, pp. 99-156, W. S. Hoar and D. J. Randall, eds. Academic Press, New York.

Hochachka, P. W., and G. N. Somero. 1973. *Biochemical Strategies in Environmental Adaptation*. W. B. Saunders, Philadelphia.

Hochachka, P. W., and T. Mustafa. 1972. Invertebrate facultative anaerobiosis. Science, *178*: 1056-1060. Charts reproduced by permission of copyright owner, American Association for the Advancement of Science, 1972.

Jackson, D. C. 1968. Metabolic depression and oxygen depletion in the diving turtle. J. Applied Physiol., *24*: 503-509.

Kemp, R. G. 1971. Rabbit liver phosphofructokinase. J. Biol. Chem., *246*: 245-252.

Llorente, P., R. Marco, and A. Sols. 1970. Regulation of liver pyruvate kinase and the PEP crossroads. European J. Biochem., *13*: 45-54.

Mangum, C. P. 1970. Respiratory physiology in annelids. Amer. Scientist, *58*: 641-647.

Mansour, T. E. 1970. Kinetic and physical properties of phosphofructokinase. Adv. Enzyme Regulation, *8*: 37-51.

Markert, C. L., and Y. Masui. 1969. Lactate dehydrogenase isoenzymes of the penguin. J. Exp. Zool., *172*: 121-146.

Moon, T. W., T. Mustafa, and P. W. Hochachka. 1971. The adaptation of enzymes to pressure. II. A comparison of muscle pyruvate kinases from surface and midwater fishes with the homologous enzyme from an off-shore benthic species. Amer. Zoologist, *11*: 491-502.

Mustafa, T., T. W. Moon, and P. W. Hochachka. 1971. Effects of pressure and temperature on the catalytic and regulatory properties of muscle pyruvate kinase from an off-shore benthic fish. Amer. Zoologist, *11*: 451-466.

Saz, H. J. 1971. Facultative anaerobiosis in the invertebrates: pathways and control systems. Amer. Zoologist, *11*: 125-135.

Scrutton, M. C., and M. F. Utter. 1968. The regulation of glycolysis and gluconeogenesis in animal tissues. Annual Rev. Biochemistry, *37*: 249-302.

Somero, G. N. 1969. Pyruvate kinase variants of the Alaskan king crab. Evidence for a temperature-dependent interconversion between two forms having distinct and adaptative kinetic properties. Biochem. J., *114*: 237-241.

One Hundred Years of Pacific Oceanography

Joel W. Hedgpeth
Marine Science Center
Newport, Oregon

OCEANOGRAPHY, or the scientific study of the sea, by whatever name, did not begin exactly with the Voyage of the *Challenger*, although the first station of that voyage, in the Bay of Biscay on December 30, 1872, has conventionally been cited as the birthday of oceanography (fig. 1). This date is academic for the Pacific in any event since the *Challenger* did not reach Pacific waters for two years, and never reached the northeastern part which is the region of our greatest concern. But since this is the year 1972, which is being celebrated as the centennial year of the *Challenger* Expedition, it seems appropriate to remind ourselves of the event. There will be many more words said on the subject before the year is over, suitably embellished with a haggis, no doubt, at the grand finale in Edinburgh. (See the postscript added in proof.)

During the century and a half before the *Challenger* brought her scientists into the Pacific there had been many noteworthy scientific travelers, although they did not have much to do with oceanography as we think of it. For the most part these adventurous naturalists were assigned to exploring or surveying vessels, and they did not have the best of luck. Georg Steller, who sailed the Bering Sea and the Aleutians with Vitus Bering and was the only naturalist to see the great sea cow alive, died in 1746 on a confused shuttling back and forth between St. Petersburg and Siberia that ensued from some sort of official investigation of the conduct of the expedition; Cook's naturalist on the *Resolution*, William Anderson, died off Alaska in 1778 and of course Cook himself never saw England again; and La Martinière, naturalist

Figure 1.   H.M.S. *Challenger,* St. Thomas, B.W.I., March 1873.

of the *Astrolabe,* was lost along with La Pérouse, when both ships of the expedition were wrecked on a reef at Vanikoro in 1788. It is not certain who survived the wreck, since it was not located until 1827, some 39 years later (Discombe and Anthonioz, 1960). More prosaically Robert Kennicott, naturalist of our first exploring expedition to Alaska, died of a heart attack (apparently) on the banks of the Yukon; the naturalists of the Wilkes Expedition (the United States Exploring Expedition of 1838-1842) had better luck: they all got home, but their reports were mired in bureaucratic contention and some were never published, although the collections became the nucleus for the U.S. National Museum. In contrast to all these casualties and difficulties, one may remember the pleasant voyage of the *Rurik* southward along the coast to San Francisco with Kotzebue, Eschscholtz, and Adalbert von Chamisso in 1816. Chamisso, a poet at heart, but acting as botanist of the cruise, made the first observation of the alternation of generations in pelagic ascidians, an auspicious start for marine biology in our part of the world.

When the *Challenger* started on her cruise one of the charges from the Royal Society was to ascertain the true nature and extent of *Bathybius,* that primordial Urschleim that some eminent biologists of the day thought might be the origin of life itself (fig. 2). It was described by Thomas Henry Huxley from deep sea samples, who named it *Bathybius haeckeli* in honor of Ernst Haeckel. *Bathybius* was "undiscovered" by the ship's chemist, J. Y. Buchanan, who determined that it was a sort of amorphous colloidal gloop caused by mixing bot-

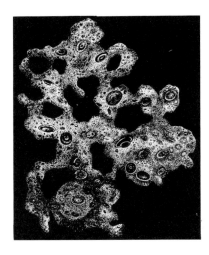

Figure 2. *Bathybius haeckeli.* From C. Wyville Thomson, *Depths of the Sea* (1873).

tom mud with impure, sulfate-rich alcohol. As the ill-fated Rudolf von Willemoes-Suhm (1877, p. 157) put it in a letter to Professor Kuppfer: "Wir sind Alle der Ueberzeugung, dass *Bathybius* nur in Spiritusflaschen lebt, niedergeschlagenes, coagulirtes Eiweiss." It would appear that in the manner of communicating his great discovery to the world (by letters from Japan) that Buchanan rather laid it on, to the discomfiture of the naturalists. He informed his chemist colleagues at home how to make *Bathybius* as a sort of parlor trick before Huxley got the word. Perhaps as a result of these and similar tactics Buchanan never fared well with the *Challenger* people in Edinburgh after the expedition was over.[1] This episode also suggests that the dichotomy between the "hard" and "soft" scientists and their rivalry for ship time was also a legacy of the *Challenger* Expedition.

The original cruise plan of the *Challenger* included a line across the north Pacific to British Columbia, but this plan was revised, evidently in Hong Kong, when her Captain, G. S. Nares, was reassigned to an Arctic expedition. Also, the U.S.S. *Tuscarora* was making a survey of the north Pacific for a cable route to the Orient at the time and was considered to be making an important contribution to oceanography. So, after proceeding eastward from Japan the *Challenger* turned due south to the Hawaiian Islands and from thence via various South Sea Islands to Juan Fernandez and Chile. The ship left the Pacific via the Straits of Magellan in January 1876. When at sea south of the Hawaiian Islands the *Challenger* lost one of the most promising

---

[1] See G. E. R. Deacon, 1968, footnote; and Buchanan's account of the episode on p. ix of his Scientific Papers, Vol. I (1913); also Buchanan, 1895.

Figure 3. Rudolf von Willemoes-
Suhm, who died aboard H.M.S.
Challenger at sea in 1875.

naturalists of the generation, Rudolf von Willemoes-Suhm, who died
of erysipelas (fig. 3). His name is remembered as a genus of Crus-
tacea. In the old tradition of the Greeks, who often placed markers on
land for those lost at sea, Willemoes-Suhm's messmates placed a monument in the cemetery of their comrade's home town in Germany. It
would have been appropriate, perhaps, had his monument borne that
inscription from the Greek anthology:

> No dust have I got to cover me,
>   My grave no man may show;
> My tomb is this unending sea
>   And I lie far below.   (E. A. Robinson).

After the departure of the *Challenger* from the Pacific in 1876
there was a gap of some years in scientific activity until the cruise of
the old corvette *Vitiaz* under the Russian scientist, Admiral S. O. Makaroff. Although this cruise, which lasted from 1886 to 1889, is not so
generally remembered, its principal contributions were to the physical
oceanography of the north Pacific. Admiral Makaroff was a scientist of
considerable competence, who apparently was the first to develop the
idea of summarizing surface temperature data by one degree squares,
and he recognized the Antarctic origin of the deeper waters of the Pacific (Soloviev, 1968).[2] It was in physical oceanography, incidentally,

---

[2] According to Soloviev's interesting account of Admiral Makaroff's work,
he was a stickler for accurate data and one of his working maxims was: "One
careless observation will spoil 100 good ones."

that the *Challenger* was weakest (G. E. R. Deacon, 1968); according to H. Burstyn (1968), the opportunity to develop modern dynamic oceanography was missed by the *Challenger*. Perhaps this is just a bit unfair, for the sort of person who might have contributed to this development was not aboard; Lord Kelvin, for example, had little to do with the cruise other than contribute his ideas on sounding machines. Also, it took time (and data) to work out these things, and there may be some connection between the development of theoretical physical oceanography in Scandinavia and the long winter nights. I remember H. U. Sverdrup telling me that the principal advantage of a long Arctic cruise was that being frozen in the ice for a few months gave you time to think.

At the time the *Vitiaz* was still working in the northwestern Pacific, the United States Fish Commission Steamer *Albatross* was transferred to Pacific waters, with San Francisco as its home port (fig. 4). The *Albatross* was the first vessel specifically built from the keel up for research; she was launched in 1882 and completed her first cruises in 1883 (Hedgpeth and Schmitt, 1945; Hedgpeth, 1947). After five years of research in Atlantic and Caribbean waters, primarily collecting specimens (some of which are still unstudied), she was sent around South America to the Pacific. She arrived in San Francisco on May 11, 1888, after sailing via the Straits of Magellan. Originally the plan was to occupy an extensive series of dredging stations en route to San Francisco, but a cholera epidemic in Chile prevented her from putting

Figure 4. The United States Fisheries Commission steamer *Albatross* (Hedgpeth, 1947).

to port for coal. A few stations, in somewhat similar localities to those of the *Challenger,* were occupied in southern Chilean waters before the change in cruise plans. For most of the time she was in Pacific waters, the *Albatross* was a fisheries research vessel, making an almost yearly cruise to Alaska on salmon and halibut investigations, working on the Pribilof Islands fur-seal patrol, or making ethnological investigations. Often she occupied dredging stations between San Francisco and Alaska on these assignments, and much of our knowledge of the deeper fauna of the northeastern Pacific has come from the collections of the *Albatross.* In the generation before us, almost every marine biologist or naturalist of prominence had some association with the *Albatross.* The position of Naturalist of the *Albatross* (whether on board or in drydock) was held by many people who later became eminent; the last person to hold this title, when the *Albatross* was tied up in Boston Harbor after being returned to the Atlantic coast, was Paul S. Galtsoff.

The greatest naturalist of the *Albatross,* however, never held that title. He could have built his own research vessel had he wanted to, but instead he made an arrangement with the federal government to pay the ship's bills for several cruises in the South Pacific in 1891, 1900, and 1904-1905. This was, of course, Alexander Agassiz, son of Louis Agassiz. It is said that he became a successful mining engineer, one of the developers of the great copper deposits in Michigan, in part as a reaction to the inadequate financial support given his father at Harvard. In any event, his heart was always with marine zoology and when he had to make the choice of dividing his efforts he withdrew from the Corporation of Harvard College in order to continue as a director of the Calumet mine. This provided him with the wherewithal to carry on his own scientific work. Yet, during nine busy years, "which Agassiz devoted to the executive work of the mine, the Museum, and the University, to say nothing of his enforced winter absences in search of health, his writings number no less than fifty-nine titles. While many of these were short articles, some of the more important publications were "Three Cruises of the *Blake,*" "*Blake* Echini," "Coral Reefs of the Hawaiian Islands," and a number of papers on the embroyology and development of bony fishes (G. R. Agassiz, 1913).

Agassiz's first *Albatross* cruise, from February to April of 1891, was from Panama to the Galapagos and up into the Gulf of California as far as Guaymas. His main interest on this cruise was the relationship between the marine fauna on the Pacific side of the Isthmus as contrasted with that of the Caribbean and Gulf of Mexico, a matter which has now become of topical interest because of proposals for a sea-level canal between the Atlantic and Pacific. During this cruise he tried closing nets for plankton sampling because he realized the drawbacks of open net tows. In 1899-1900 Agassiz took the *Albatross* to the Marquesas and Japan; on this cruise he made one of the deepest successful dredge hauls of record (7,632 meters), not to be surpassed until the *Galathea's* deep dredging in the Pacific trenches in 1951, but his primary interest on this cruise was coral reefs. Agassiz's third *Albatross* expedition was from Panama to Easter Island (where he passed his 69th birthday)and the Galapagos; the principal emphasis of this cruise was the Humboldt Current, as it was then known. Charles A. Kofoid, later chairman of zoology at Berkeley, was on this cruise. By all accounts Alexander Agassiz was a remarkable marine zoologist; apparently he knew at sight most of the organisms described on previous expeditions and recognized them as they were sorted on deck. He was also a remarkable person, as attested by the words of his friend Henry Adams: "He was the best we ever produced, and the only one of our generation whom I would have liked to envy. . . . We did one first-rate work when we produced him, and I do not know that, thus far, any other country has done better." (G. R. Agassiz, 1913). Agassiz died at sea on the Atlantic Ocean on Easter Sunday 1910, and is buried in Cambridge.

Inspired perhaps by the princely example of Alexander Agassiz, the railroad magnate, Edward H. Harriman, converted his hunting trip to Alaska in 1899 into a scientific expedition (Merriam, 1902). Soliciting advice from the Smithsonian Institution, he invited a number of eminent scientists and promising students to accompany his family on the two-month cruise, and two eminent and well-known nature writers, John Burroughs and John Muir, who were listed as "ornithologist and author" and "author and student of glaciers" respectively. It was a highly organized expedition, with committees for almost everything, including library, literature and art, and music and entertainment. Membership of most of these cultural committees was for the most part from the Harriman family and personal entourage. Actually the scientific committees were quite respectable, and among the younger zoologists on the cruise were Wesley R. Coe and Leon J. Cole.

The significance of this expedition rests not with its imposing guest list, but with the handsome support given by Mr. Harriman to the completion of the reports. In marine zoology, many of the reports included substantial parts of *Albatross* collections as well as material from the California Academy of Sciences. The monograph on the *Pycnogonida* by Leon J. Cole, for example, is still the most comprehensive treatment of these animals on the Pacific coast. The reports were published simultaneously, with color plates and collotypes, on elegant heavy weight (non-recycled) paper, and the whole series of ten volumes appeared in record time.[3] Unfortunately there has not been such elegant patronage since. Captain Allan Hancock's support of activities at the University of Southern California may well represent a considerably greater contribution to science, but it seems to have lacked the grand manner of the railroad baron.

Although the Harriman Alaska Reports were published after 1900, the expedition marked the end of an era, insofar as the Pacific was concerned. During the first three decades of the twentieth century remarkably little was contributed to the oceanography of the Pacific, except in its southernmost limits by several Antarctic expeditions and the early work of the *Discovery* until the non-magnetic hermaphrodite brig *Carnegie* made her seventh and last cruise in the Pacific in 1928-1929 (Paul, 1932).[4] Indeed, except for the work of the *Meteor* in the Atlantic and assorted small cruise efforts, little was being done in oceanography anywhere. Several marine biological stations were established in the 1890's and early 1900's, the first of these being the Hopkins Marine Station of Stanford University. During the first season of the Hopkins station at Monterey, in 1892, William Emerson Ritter of the University of California taught a field course and after that set out to find his own marine station. He settled upon the San Diego area,

---

[3] There seems to have been no anecdotal account of this comparatively brief expedition. Wesley R. Coe, then a young assistant professor at Yale, was on the trip and was fond of telling, during his charming anecdotage at Scripps some years back, various stories. One of these concerned a quiet morning in an Alaska harbor at low tide. Walking along the dock, Dr. Coe saw a large octopus clinging to the piling at the water line. He commandeered a skiff nearby and wrested the octopus from the pile. Quite a struggle ensued, but finally the octopus collapsed. By this time a crowd had gathered on the dock to watch the spectacle, and one of them remarked, "Well, we thought he was going to get you." I never did learn the final fate of the specimen.

[4] The loss of the *Carnegie* with her Captain by fire in Apia Harbor is a reminder of the otherwise remarkable record of research vessels. Even if one includes the two ships of La Perouse from the dangerous days of exploration by sail in uncharted waters, only four research vessels have been lost on duty. The other loss is that of the No. 5 *Kaiyô-Maru* which sailed too close to a volcano with a scientific staff of geologists. A much greater loss to oceanography was that of Townsend Cromwell and a group of colleagues in an airplane crash in 1958. The loss of the San Diego Marine Biological Laboratory schooner *Loma* on Point Loma in 1907 was a comparatively minor incident; there were no casualties and most of the equipment including the engine was salvaged. Scripps was to suffer still another ship loss when the original *E. W. Scripps* (replacement for the *Loma*) caught fire and burned on November 13, 1936. She was promptly replaced by *E. W. Scripps,* but the intention to name this new vessel (actually the yacht *Serena*) the *M. F. Maury* was objected to by the Scripps family and the name *E. W. Scripps* retained. I cannot locate it at the moment, but the letter from H. U. Sverdrup on the subject is just a bit testy in tone.

where the first formal sessions were held in the summer of 1904. This became eventually the Scripps Institution of Biological Research in 1911. To the north the laboratories at Friday Harbor were begun by Trevor Kincaid in 1903. The early plans to make this a cooperative effort were never accomplished, a'though as late as 1915 the universities of Kansas and Oregon were among those mentioned as considering some sort of joint arrangement (Ritter, 1915). Ritter's interests were broad, but the direction of the institution he founded was changed by Thomas Wayland Vaughan who succeeded him as director and had Scripps renamed as Scripps Institution of Oceanography in 1925. A name, however, does not change things of itself, and for some years there was hardly any blue water oceanography at Scripps or, for that matter, anywhere else in the world.

In 1926 at the Third Pan-Pacific Science Congress in Tokyo an international committee was organized, to be known as the Committee on the Oceanography of the Pacific, with three subcommittees, on physical and chemical oceanography, "fundamental marine biology," and fisheries technology; later a subcommittee on coral reefs was added. Thomas Wayland Vaughan was first chairman of this committee. The early endeavors of this committee resulted in various resolutions. Of some interest are several passed at the Fifth Pan-Pacific Science Congress meeting in Vancouver in 1928. Among these were one for the standardization of chemical data, replacement of the *Carnegie*, routine installation of thermographs on vessels, a plea for oceanographic studies in the region of the Galapagos, and various fishery matters. The most significant was resolution No. 16, calling attention to our general lack of information about those areas of the north Pacific Ocean between the Hawaiian Islands and the Aleutians and "within the triangle bounded by the North American coast and lines from Hawaiian Islands to San Francisco and to Panama," and calling for more study in these areas.

Some of this academic interest in oceanography undoubtedly was inspired by William Beebe's expeditions to the Galapagos in 1923 and the Galapagos and Humboldt Current in 1925 and certainly by his famous Bathysphere dives off Bermuda in the 1930's. All of these episodes were widely reported in the press, given prominent space in the *National Geographic*, and recounted in handsome books by Beebe himself (Beebe, 1924, 1926, 1934). He was the Captain Cousteau of my school generation, attracting public attention and stimulating the interest of younger people by his writings and public appearances (fig. 5). He also, through his books on the Galapagos, conveyed a somewhat erroneous impression of the islands as a paradise waiting settlement; several groups from Germany came to Floreana and their set-

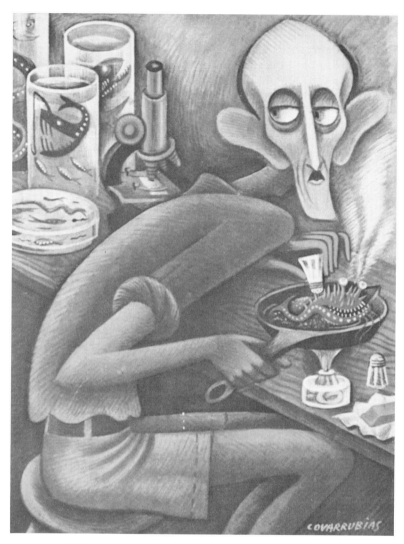

Figure 5. William Beebe, secretly frying his new species. From a caricature by Covarrubias, *Vanity Fair*, November 1933.

tlement set in motion a bizarre train of events culminating in the death of a vegetarian dentist from contaminated (poisoned?) chicken and the mysterious disappearance of the mad baroness and her entourage of paramours (Strauch, 1936; Wittmer, 1959). Captain Allan Hancock got peripherally involved because of his frequent collecting trips to the Galapagos. And, it is obvious from Chapter 27 of *Dr. Faustus,* Thomas Mann was one of Beebe's many readers. In 1936 Beebe cruised on the *Zaca* to the Gulf of California, without much contribution to science, but still another popular book (Beebe, 1938). Some years later the *Zaca* fell into the hands of Errol Flynn, who chartered it for a cruise for his zoologist father, T. Thomson Flynn, author of several papers on the Pycnogonida. Carl Hubbs went on this cruise and has a delightful story about it which unfortunately is not easily repeatable in print. But again, little serious science came out of it. Now both the Sea of Cortez and the Galapagos are of particular current interest to the oceanographers; the Galapagos has become part of the standard route of research vessels because of its position athwart the Equatorial (Cromwell) Undercurrent (see, e.g., Wooster and Hedgpeth, 1966).

Meanwhile, the National Academy of Sciences adopted a resolution in April 1927, calling for a National Academy committee to "consider the share of the United States of America in a worldwide program of oceanographic research." Among the results of this committee's work was the establishment (with Rockefeller Foundation support) of the Woods Hole Oceanographic Institution and contributions for oceanography buildings at the University of Washington and Scripps Institution. The Pacific committee continued, and at the meetings of the Sixth Pacific Science Congress, held in Berkeley, San Francisco, and Stanford in 1939, a great many papers were presented. Short versions of these papers were published in the third volume of the proceedings. The most significant work of this period, however, was the preparation of the comprehensive text, *The Oceans,* by H. U. Sverdrup, Martin W. Johnson, and Richard H. Fleming at Scripps Institution of Oceanography, which was to be published during the latter part of 1941.

But not too much was really happening in the oceans, especially in the Pacific. Despite impressive committee reports (Vaughan, 1937; Sverdrup, 1940; and Thompson, 1940), oceanography had yet to reach its eminence as part of our national effort. This was not to occur until after World War II. In 1940 there were two very different expeditions to the Gulf of California, the cruise of the *E. W. Scripps* and the *Western Flyer.* The *Western Flyer's* report, a philosophical, literary, and natural history hybrid (Steinbeck and Ricketts, 1941), was officially published on the unfortunate date of December 7, 1941, whereas the results of the Scripps cruise did not appear until 1950 (Anderson, 1950).

Nevertheless, a fair amount of information about plankton abundances, water movements, and temperatures was accumulating, but without much sense of direction or coordination. One biologist, outside the fold, attempted to synthesize what information he could find since he realized the significance of the events in the surface waters offshore to the composition and fluctuation of groupings of animals (and plants) in the tidal zone. But Ed Ricketts did not have good information—no one did—and his noble attempt to make sense of the data, as summarized in his chapter on plankton for the second edition of *Between Pacific Tides* was not a successful interpretation (Ricketts, 1948). He had hoped it would upset the professionals at Seattle and Scripps and stimulate them to some honest work. They were for the most part already stimulated, and Ricketts' effort had even less impact than he had hoped. Nevertheless he was not satisfied that he had the answers—he never was. He still hoped to be able to make a few more touches on it when the page proofs were returned, but Ricketts died just after returning galleys in April of 1948 (see Hedgpeth, 1971). Now his plankton chapter reads rather quaintly, and I thought it best to remove it from later editions of *Between Pacific Tides* and replace it with a general summary of the CalCOFI work (Hedgpeth, 1962). One of the most serious defects was Ricketts' reliance upon W. E. Allen's diatom data from Scripps pier; if nothing else, we have learned that a single point source of biological data, no matter how continuous, does not provide us with predictive information. Nevertheless, it was a noble attempt at synthesis and no one since has made a similarly whole-hearted effort to see the whole thing together or, as Ed would have put it, the "toto-picture."

By the time of Ed's unfortunate death it was obvious to all that the sardines, constituting one of the world's great fisheries, were in a bad way.[5] Nobody knew exactly why, and one of the great oceanographic programs of history was organized to study the matter. The California Cooperative Fisheries Investigations (CalCOFI) began in 1948, and by 1950 its first progress report was published. Today the annual report has become a journal, volumes of data have been accumulated, dozens of oceanographers have received their Ph.D.'s, and the work still goes on. The sardines have not come back, however, and perhaps they never will, at least in our time. But for the last quarter of a century we have had unprecedented research in biological oceanography in the northeastern Pacific. We have gained much in understand-

---

[5] And now, south of the Equator, there are signs that the Peruvian anchovy fishery is in trouble. If this fishery also goes into an irrevocable decline, it will have noticeable effect on the red and white striped bazaars of Colonel Sanders fried chicken, for the bulk of such clupeid fisheries goes into fish meal for livestock food. And there will be another big oceanographic push—perhaps a PerCOFI?

ing of the physical as well as biological processes of the sea from this research, and alongside this record the febrile, draconically precipitate, and mission-oriented efforts of the Sea Grant Program seem more like landlubbers' impertinence.

During those euphoric postwar (and post-sardine) years at Scripps the magnitude of the plankton problem was becoming apparent to physical oceanographers and engineers, who hoped for a philosopher's stone or at least a short cut across an apparently hopeless impasse. Obviously, different species ought to have different physical properties—density, perhaps—so perhaps they could be separated physically. One imaginative engineer constructed an elaborate apparatus, somewhat on the principle of the old dairy farm cream separator, with numerous lucite pigeon holes into which the different plankters would obligingly arrange themselves given an appropriate amount of agitation. The apparatus had numerous small motors and other plunder from the surplus bins and for years kept scavengers of the second order happy. It also had three large conical udderlike structures along its venter, suggestive of some primordial theromorph. Alas, it did not work and for years was in the way of successive occupants of the room. Finally, and appropriately, perhaps, it found shelter in the office of a glamorous scientist from another land who thought that the solution might lie in the biochemical properties of the species, that is, each species should have a different autograph when squashed in a sheet of chromatograph paper, and he had the sad bones of the plankton separator hauled away to the dump. He too has gone his way with his sheets of paper and Monel metal bins and the plankton problem still remains.

Perhaps more imagination was really needed, at least some new approaches, and for a while "perspectives" was the word. A fine international conference was held in La Jolla in 1956 to produce some of these perspectives; the Madison Avenue approach of buzz sessions was tried. The meeting produced some interesting papers, lively discussions, and made some friendships, as all conferences do, but it was undoubtedly disconcerting to those who hoped that traditional systematic approaches could be bypassed to hear, as the first fruit of the intellectual ferment, a recommendation for strengthening systematics and encouraging more taxonomists (in the summary this recommendation was politely placed number 6, although it was stated that systematics "is the basis on which the rest of biology builds." Buzzati-Traverso, 1960). Things do move slowly, even if it may be plain to a biologist that most of our interest in and concern about the oceans would not exist if it were an abiotic solution of miscellaneous chemicals. At least it can now be said that we have progressed, in public understanding,

beyond the sarcastic commentator on the *Challenger* Expedition, who
wrote:

> The first volume recording the adventures of the *Challenger* yacht-
> ing trip is now out, and the other fifty-nine will be ready in less than a
> century. Everybody knows that Mr. Lowe sent a man-of-war away
> laden with Professors, and that these learned individuals amused them-
> selves for four years. They played with thermometers, they fished at all
> depths from two feet to three miles; they brought up bucketfuls of stuff
> from the deep sea bottom; and they all pottered about and imagined they
> were furthering the grand Cause of Science. . . . The whole business
> has cost two hundred thousand pounds; and in return for this sum we
> have got one lumbering volume of statistics, and a complete set of
> squabbles which are going on briskly wherever two or three philisophers
> are gathered together. I believe the expedition discovered one new spe-
> cies of shrimp, but I am not quite sure.   (M. Deacon, 1971, p. 371-372.)

No one now would write so disrespectfully, or at least with such
misunderstanding, of oceanography, even though the massive funding
of the early postwar years is over. In those times the Navy had as
much interest in basic science as in Mission Oriented Research, was
generous with funds and surplus ships, and indeed the worldwide climate for scientific work at sea was in its halcyon days. Everywhere students were encouraged to become oceanographers, and some of them
are contributors to this volume. For the Scandinavians it was the time
to reinstate the old tradition of round-the-world cruises: the Swedes

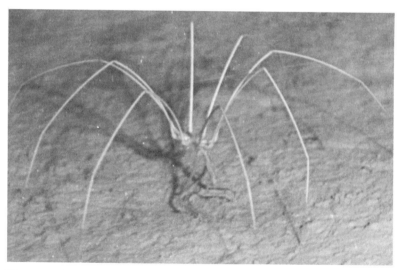

Figure 6. *Colossendeis colossea,* a significant deep sea pycnogonid. Photograph
courtesy Robert Hessler.

sent their *Albatross* around the world in 1947-1948 and the Danes followed suit with *Galathea* in 1950-1951. The principal emphasis of the *Albatross* was physical oceanography and marine geology, especially the collection of very long sediment cores, although her biological collections were not entirely without significance, especially the pycnogonid (fig. 6). (Hedgpeth, 1954). The *Galathea* was primarily a biological ship, bringing up many new and strange creatures from great depths, including the now famous primitive mollusc *Neopilina*. To the great distress of geologists, one movie taken aboard showed the biologists carefully removing organisms from stones that had been in the ocean's deepest holes before casting the stones back into the sea. Lesser voyages, sometimes even more parochial in mission, followed suit and the circumstances of their going and coming can still stir up nostalgia for the old days at Scripps when Bascom's bubbles rose in steps and certain eminent worthies were known as Mattresshead and Blanketfoot. Such times are gone along with their leading spirit, who has become beached on the hemocyaninhued shores of Harvard where he attempts to follow the humanitarian example of Fridtjof Nansen and is categorized for his pains by Garrett Hardin as a "cornucopist." (For an anecdotal account of some of these eminences, see Behrman, 1969.)

The Russians entered the oceanological lists in the 1950's with repeated expeditions to the North Pacific trenches, and the Japanese also developed a program which is now of the first order; both countries mounted Pacific expeditions as far south as the Antarctic. The incentive for all this research was primarily to gather information on which to estimate the potential yield of the ocean and, hopefully, to predict fisheries stocks. Certainly the committees of the 1920's would be gratified by this international assault in behalf of all aspects of oceanography.

In addition to the stimuli from military and economic needs, our tampering with the atom and the development of nuclear powered submarines required the services of many oceanographers. It was not always plain what these fellows were up to at the time, leaving for months on end on code worded missions and quietly returning. One series of studies was undertaken in response to the use of the Columbia River as an atomic effluent system for the Hanford works; rencetly much of this information has been gathered together (Pruter and Alverson, 1972) and is the nearest thing we have to a compilation of regional oceanography in English. There is now a wealth of information in the various CalCOFI reports, including the atlasses of temperature, geostrophic heights, copepods, euphausiids, chaetognaths, and vast accumulations of data from the last three decades of activity (including such north Pacific synoptic efforts as NORPAC) awaiting

Figure 7. Four late, great figures of Pacific marine biology. Wilbert McLeod Chapman, 1910-1970 (upper left) ; Milner Baily Schaeffer, 1912-1970 (upper right) ; Oscar Elton Sette, 1900-1972 (lower left) ; and Lev Alexsandrovich Zenkevich, 1889-1970 (lower right).

synthesis. Everyone hopes that someone talented enough to draw all this together will step forward and be recognized—and, hopefully, be supported in his synthetic effort. As always, we seem to lack literate, synthetic oceanographers. I must say that my optimism that such a person will eventually materialize is tempered by my experience with graduate seminars, where the simple requirement that the report is to be written in "acceptable English" automatically cuts the enrollment in half.

Oceanographers, especially of the biological version, are now a conspicuous part of the scientific population all around the Pacific rim; to mention the most prominent active practitioners would be to betray personal prejudice and (sometimes) inadvertently slight others equally worthy. But we must note with sadness that the giants of our time are beginning to fall by the wayside. In 1970, within a few months of each other, Wilbert McLeod Chapman, Lev Alexandrovitch Zenkevitch, and Milner B. Schaefer, all of them influential and prolific contributors to the oceanography and scientific activity of the north Pacific, died; in 1972 Oscar Elton Sette, an influence behind the scenes for most of CalCOFI's years, joined them (fig. 7). Men like these cannot be replaced; those who rise from the ranks will fill other positions or occupy other niches. But their eminence would not have been possible without the rest of us still carrying on. As Beaglehole (1934) said in his final peroration of the unknown but no less gallant explorers of the Pacific, so may we say of the many hard-working scientific sailors, systematists, monographers, museum drudges, and preoccupied professors (and, in our day, contract workers) who have brought us to our present knowledge of the biology of the Pacific:

> Yet one gazes on the waters of the Pacific today, that ocean which washes so many continents and knows alike the icy battlements of both poles, beating in long surf on so many tropic shores, not without memory of the ships of other voyagers nor their ambition; for in them, however they differed in accomplishment or intent, lay fundamentally that unity of spirit which in some sort gives constancy and wholeness to the inconstant and fragmentary lives of men: of them it may be said that their effort was not without result, and in the enlargement of the knowledge with which men contemplate this globe they have their allotted place.

Postscript
(Added in proof)

The prophecy that haggis would be served at Edinburgh was not fulfilled (although another product of Scotland was in satisfactory supply), but there have indeed been many words in memory of the *Chal-*

lenger. The proceedings of the *Challenger* Expedition Centenary occupy volumes 72 and 73 of The Proceedings of the Royal Society of Edinburgh, published of course in 1972. Contributions of special interest to the history of oceanography in the north Pacific are by Robert W. Sandilands on hydrographic charting and oceanography in British Columbia (Vol. 73, pp. 75-83) and Mitchitaka Uda on fisheries oceanography in Japan (*idem,* 391-398). The September 1972 issue of the Geographical Magazine contained several articles about the *Challenger* Expedition; the most noteworthy of these in our context is by David Stoddart, "Buchanan—the forgotten apostle" (pp. 858-862). A discussion of the origins of the *Challenger* Expedition by Herbert S. Bailey Jr. appeared as the lead article in the September-October 1972 issue of American Scientist (pp. 550-560). We may expect such observances perhaps into 1976: Oceans magazine recognized the centenary in its last issue (no. 6) for 1973 with an article by Phillip Drennon Thomas (pp. 41-45).

Literature Cited

Agassiz, G. R., ed. 1913. *Letters and Recollections of Alexander Agassiz, with a Sketch of his Life and Work.* Houghton Mifflin, Boston.
Anderson, Charles A., ed. 1950. 1940 *E. W. Scripps* Cruise to the Gulf of California. Geol. Soc. Amer., Mem. 40.
Beaglehole, J. C. 1934. *The Exploration of the Pacific.* A. & C. Black, London.
Beebe, William. 1924. *Galapagos, World's End.* G. P. Putnam's Sons, New York and London.
Beebe, William. 1926. *The Arcturus Adventure. An Account of the New York Zoological Society's First Oceanographic Expedition.* G. P. Putnam's Sons, New York.
Beebe, William. 1934. *Half Mile Down.* Harcourt, Brace & Co., New York.
Beebe, William. 1938. *Zaca Venture.* Harcourt, Brace & Co., New York.
Behrman, Daniel. 1969. *The New World of the Oceans. Men and Oceanography.* Little Brown & Co., Boston.
Buchanan, J. Y. 1895. A retrospect of oceanography in the twenty years before 1895. Report Sixth International Geographical Congress; Reprinted in Accounts Rendered of Work Done and Things Seen, Cambridge, 1919.
Buchanan, J. Y. 1913. Scientific Papers. Cambridge, 1913. (Original paginations).
Burstyn, Harold L. 1968. Science and government in the nineteenth century: The *Challenger* Expedition and its report. Bull. Inst. Oceanogr., Special no. 2: 603-613.
Buzzati-Traverso, A. A., ed. 1960. *Perspectives in Marine Biology.* University of California Press, Berkeley.
Deacon, G. E. R. 1968. Early scientific studies of the Antarctic Ocean. Bull. Inst. Oceanogr., Special no. 2: 269-279.
Deacon, Margaret. 1971. *Scientists and the Sea, 1650-1900. A Study of Marine Science.* Academic Press, New York.
Discombe, Reece, and Pierre Anthonioz. 1960. Voyage to Vanikoro. Pacific Discovery, Jan.-Feb. 1960, *13*(1): 4-15.

Hedgpeth, Joel W. 1945. The United States Fish Commission Steamer *Albatross*. With an appendix by Waldo L. Schmitt. American Neptune, *5*(1): 5-26.

Hedgpeth, Joel W. 1946. The Voyage of the *Challenger*. Scientific Monthly, *63:* 194-202.

Hedgpeth, Joel W. 1947. The steamer *Albatross*. Scientific Monthly, *65*(1): 17-22.

Hedgpeth, Joel W. 1954. Reports on the dredging results of the Scripps Institution of Oceanography Trans-Pacific Expedition, July-December, 1953. I. The Pycnogonida. Systematic Zoology, *3*(4): 147.

Hedgpeth, Joel W. 1962. Beyond the tides: the uncertain sea. In *Between Pacific Tides,* 3rd ed., rev. by E. F. Ricketts and Jack Calvin, pp. 376-406. Stanford University Press, Stanford.

Hedgpeth, Joel W. 1971. Philosophy on Cannery Row. In *John Steinbeck. The Man and His Work,* pp. 89-129, Richard Astro and Tetsumaro Hayashi, eds. Oregon State University Press, Corvallis.

Hedgpeth, Joel W. 1973. Genesis of Sea of Cortez. The Steinbeck Quarterly, *6*(3): 74-80.

Merriam, C. Hart, ed. 1902. Alaska. Harriman Alaska Expedition, Vol. 1. Doubleday, Page, New York.

Murray, John, ed. 1885. *Narrative,* Vol. I, *Report of the Scientific Results of the Voyage of H. M. S. Challenger.*

Paul, J. Harland. 1932. *The Last Cruise of the Carnegie.* Williams & Wilkins, Baltimore.

Pruter, A. T., and D. L. Alverson, eds. 1972. *The Columbia River Estuary and Adjacent Ocean Waters. Bioenvironmental Studies.* University of Washington Press, Seattle.

Ricketts, Edward F. 1948. Marine plankton of the Pacific Coast. In *Between Pacific Tides,* 3rd ed., rev. by E. F. Ricketts and Jack Calvin, pp. 255-283. Stanford University Press, Stanford.

Ritter, William Emerson. 1915. The biological laboratories of the Pacific Coast. Pop. Science Monthly, March 1915: 223-230.

Soloviev, A. I. 1968. S. O. Makarov and the significance of his research in oceanography. Bull. Inst. Oceanogr., Special no. 2: 615-625.

Steinbeck, John, and Edward F. Ricketts. 1941. *Sea of Cortez: a Leisurely Journal of Travel and Research.* Viking, New York.

Strauch, Dore. 1936. *Satan Came to Eden.* Harper & Brothers, New York.

Sverdrup, H. U. 1940. Activities of the Scripps Institution of Oceanography, La Jolla, California. Proc. Sixth Pacific Science Congr., *3:* 114-123.

Thompson, Thomas G. 1940. Report of the Standing Committee on the Oceanography of the Pacific. Sixth Pacific Science Contr., *3:* 5-12.

Thomson, C. Wyville. 1873. *The Depths of the Sea.* MacMillan & Co., London.

Vaughan, Thomas Wayland. 1937. *International Aspects of Oceanography.* National Academy of Sciences, Washington, D.C.

Wittmer, Margret. 1959. *Postlagernd Floreana. Robinsonfrau auf den Galapagos-Inseln.* Buchergilde Gutenberg, Frankfurt am Main.

Willemoes-Suhm, Rudolf von. 1877. *Challenger-Briefe von Rudolf v. Willemoes-Suhm Dr. Phil. 1872-1875 nach dem Tode des Verfassers herausgegeben von seiner Mutter.* Wilhelm Engelmann, Leipzig.

Wooster, Warren S., and Joel W. Hedgpeth. 1966. The Oceanographic Setting of the Galapagos. In *The Galapagos.* Robert I. Bowman, ed., pp. 100-107. University of California Press, Berkeley.

Appendix

Thirty-third Annual Biology Colloquium

Theme: The Biology of the Oceanic Pacific

Dates: April 21-22, 1972

Place: Oregon State University, Corvallis, Oregon

Colloquium Committee: Charles B. Miller, program; John V. Byrne, sponsorship; Robert Holton, banquet; James McCauley, advertising.

Standing Committee for the Biology Colloquium: J. Ralph Shay, chairman; Robert R. Becker, J. C. Dilworth, Robert Doerge, Ernst Dornfeld, Paul Elliker, Betty Hawthorne, Charles B. Miller, Knud Swenson, John Wiens, David Willis.

Colloquium Speakers (current addresses, fall, 1973):

John A. McGowan, Scripps Institution of Oceanography, La Jolla, California, leader

Timothy R. Parsons, Institute of Oceanography, University of British Columbia, Vancouver, B.C.

Bruce W. Frost, Department of Oceanography, University of Washington, Seattle, Washington

Robert R. Hessler, Scripps Institution of Oceanography, La Jolla, California

Peter W. Hochachka, Department of Zoology, University of British Columbia, Vancouver, B.C.

Brian J. Rothschild, National Marine Fisheries Service, La Jolla, California

Joel W. Hedgpeth, Marine Science Center, Newport, Oregon

Sponsorship:

Sea Grant Program, Oregon State University
Research Council, Oregon State University
School of Science, Oregon State University
Agricultural Experiment Station, Oregon State University
Graduate Council, Oregon State University
Sigma Xi
Phi Kappa Phi

(From Murray, 1885)

DATE DUE

MY 12 '82			